AS/A-LEVEL YEARS 1 AND 2

STUDENT GUIDE

AQA

Physics

Practical assessment

Graham George and Kevin Lawrence

Series editor: Graham George

HODDER
EDUCATION
AN HACHETTE UK COMPANY

Hodder Education, an Hachette UK company, Blenheim Court, George Street, Banbury, Oxfordshire OX16 5BH

Orders

Bookpoint Ltd, 130 Park Drive, Milton Park, Abingdon, Oxfordshire OX14 4SE

tel: 01235 827827

fax: 01235 400401

e-mail: education@bookpoint.co.uk

Lines are open 9.00 a.m.–5.00 p.m., Monday to Saturday, with a 24-hour message answering service. You can also order through the Hodder Education website: www.hoddereducation.co.uk

© Graham George and Kevin Lawrence 2017

ISBN 978-1-4718-8515-0

First printed 2017

Impression number 5 4

Year 2020 2019

This guide has been written specifically to support students preparing for the AQA A-level Physics examinations. The content has been neither approved nor endorsed by AQA and remains the sole responsibility of the authors.

Cover photo: Peter Hermes Furian/Fotolia; page 58, GIPhotostock/SPL.

Typeset by Integra Software Services Pvt. Ltd, Pondicherry, India

Printed in India

Hachette UK's policy is to use papers that are natural, renewable and recyclable products and made from wood grown in sustainable forests. The logging and manufacturing processes are expected to conform to the environmental regulations of the country of origin.

Contents

Content Guidance

Questions & Answers

■ Getting the most from this book

Knowledge check

Rapid-fire questions throughout the Content Guidance section to check your understanding.

Knowledge check answers

Turn to the back of the book for the Knowledge check answers.

Exam tips

Advice on key points in the text to help you learn and recall content, avoid pitfalls, and polish your exam technique in order to boost your grade.

Commentary on the questions

Tips on what you need to do to gain full marks, indicated by the icon **e**

Sample student answers

Practise the questions, then look at the student answers that follow.

Exam-style questions

Commentary on sample student answers

Read the comments (preceded by the icon **e**) showing how many marks each answer would be awarded in the exam and exactly where marks are gained or lost.

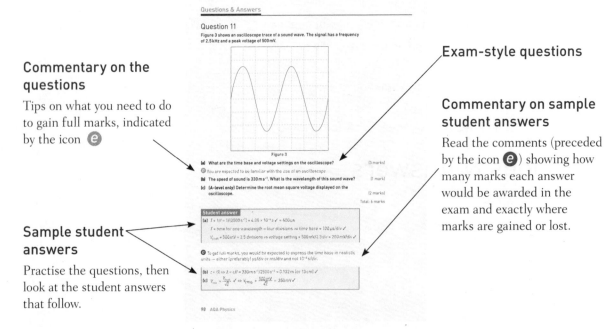

■ About this guide

This guide is one of a series covering the AQA specification for AS and A-level physics. It offers advice for the effective development of practical skills. Its aim is to help you understand and put into practice the physics — it is not intended as a shopping list, enabling you to cram for the examination. The guide has four sections:

- The **Maths and Units** section incorporates the mathematical skills that you need for the AS and A-level examination, together with a reminder of the importance of units. For A-level you need more advanced mathematical skills, which are highlighted in **bold**. This section provides the basis of all calculations that you will be required to undertake and that is why it is the first section in the book. Without these mathematical skills, you will not be able to analyse your data and subsequently draw conclusions about your experiment. You are advised to work through this section to remind yourself of the essential mathematics and put your knowledge to the test by attempting the questions as you go along. You may also find it useful to refer back to this section when you are trying questions later in the book.

- The **Practical Skills** section develops the essential skills needed for successful practical work, based on the practical criteria set out in the specification. It will help you become a confident practical physicist, able to use a variety of apparatus and techniques to collect data, which can then be analysed and used to draw valid conclusions.

- The **Required Practicals** section reflects the AQA requirement that you should acquire competence and confidence in a variety of practical, mathematical and problem-solving skills and in handling apparatus competently and safely. Each of the 12 required practicals is discussed in detail, with worked examples based on experimental data.

- The **Questions & Answers** section pulls together the other three sections through a range of questions based on the type of question that will be asked in examination papers to test practical knowledge, particularly in paper 2 at AS and paper 3 at A-level. It does *not* in itself replicate a complete examination paper, as it contains only questions that test practical skills. Answers are provided and common errors made by students are also highlighted, so that you, hopefully, do not make the same mistakes.

The purpose of this book is to help you answer A-level papers, but do not forget that what you are doing is learning physics. The development of an understanding of physics can only evolve with experience, which means time spent thinking about physics, working with it and solving problems. This book provides you with a platform for doing this.

If you are reading this, you are clearly determined to do well in your examinations. If you try all the knowledge checks, the worked examples and the questions at the end *before* looking at the answers, you will begin to think for yourself and develop the necessary techniques for answering examination questions. As the answers to the worked examples are an integral part of the learning process, they appear immediately after the questions. You are recommended to cover up the answers to prevent 'cheating' — after all, if you 'cheat' by looking up the answers first, you are only cheating yourself.

Thus prepared, you will be able to approach the examination with confidence. Good luck, and remember that physics is fun!

Content Guidance

■ Maths and units

Throughout your A-level course, you will make use of a range of mathematical knowledge and techniques. Some of these, such as basic algebra, will be used regularly in many different topics. Others, such as logarithms, will only be needed in a few topics. In this section, we shall cover the range of mathematics required during your course, with a particular focus on how they are used in practical work. Those statements in **bold** will be assessed in the full A-level course only.

Arithmetic and units

Correctly interpreting numerical information is an important first step in many practical calculations in physics.

Throughout your studies you will need to:

- recognise and make use of appropriate units in calculations
- recognise and use expressions in decimal and standard form
- use ratios, fractions and percentages
- estimate results
- use calculators to find and use power functions, **exponential** and **logarithmic functions**
- use calculators to handle $\sin x$, $\cos x$ and $\tan x$ when x is expressed in degrees or radians

Physical measurements are expressed in standard (SI) units (Table 1). These have been agreed internationally so that the same value and unit for each is used in different countries.

Table 1 Standard SI base quantities and units

Quantity	Unit	Symbol
length	metre	m
mass	kilogram	kg
time	second	s
electric current	ampere	A
temperature	kelvin	K
amount of substance	mole	mol

These base SI units are combined to give units for other quantities. The units of many commonly used quantities are given their own names. For example, force is measured in newtons (N) but this can be expressed as $\text{kg}\,\text{m}\,\text{s}^{-2}$; pressure is measured in pascals (Pa), which is the same as $\text{kg}\,\text{m}^{-1}\,\text{s}^{-2}$.

> **Exam tip**
>
> Always look carefully at the units in any numerical data. It is often helpful to convert units into standard units such as metres (m), rather than leave them as millimetres (mm).

> **Knowledge check 1**
>
> Give the following units in terms of base SI units: **a** joule, J, **b** coulomb, C, **c** ohm, Ω, **d** watt, W.

For very small or very large measurements, prefixes are used. Table 2 gives some of the common prefixes used in physics.

Table 2 Standard SI prefixes

Name and symbol	Factor of 10	Name and symbol	Factor of 10
centi, c	10^{-2}		
milli, m	10^{-3}	kilo, k	10^{3}
micro, μ	10^{-6}	mega, M	10^{6}
nano, n	10^{-9}	giga, G	10^{9}
pico, p	10^{-12}	tera, T	10^{12}
femto, f	10^{-15}		

Knowledge check 2

A student is investigating the density of a length of wooden dowelling. She measures the diameter of the wood to be 12.7 mm and the length of the dowelling as 80.0 cm. She calculates that the area of cross-section of the dowelling is 127 mm² and that its volume is 102 cm³. Convert these four measurements into m, m² and m³, expressing your answers in standard form if appropriate.

Worked example

(**A-level only**) A student is measuring the gravitational attraction between two masses, M and m. Estimate what happens to the gravitational force between the masses when the distance separating them is halved.

Answer

For the original force:

$$F_1 = \frac{GMm}{r^2}$$

When the distance is halved:

$$F_2 = \frac{GMm}{\left(\frac{r}{2}\right)^2} = \frac{GMm}{\frac{r^2}{4}} = \frac{4GMm}{r^2}$$

But $F_1 = \frac{GMm}{r^2}$ so $F_2 = 4F_1$.

Knowledge check 4

The refractive index of olive oil is 1.47 and the refractive index of water is 1.33. Calculate the following ratio: speed of light in water (c_w) : speed of light in olive oil (c_o).

Sometimes, when solving real-world problems in physics, you will not have all the necessary information that you need. It is helpful to be able to make a reasonable estimate for the answer to the problem. This can be done using order-of-magnitude figures for the variables in the problem. The physicist Enrico Fermi was able to calculate

Knowledge check 3

The light from a laser has a wavelength of 643 nm and a diffraction grating has a distance between slits of 3.3 μm. Express these values in m and in standard form.

Exam tip

Always check whether the values you are given in an exam question have a prefix. Convert these into standard form before you attempt to answer the question.

Exam tip

Multiple-choice questions will often require you to make use of ratios to decide how changing one of the variables in an equation will affect another variable.

Knowledge check 5

a If X is 120, what is 40% of X?

b If 20% of X is 60, what is X?

c If 25% of X is 2.0 × 10^3 what is X?

the approximate strength of the first nuclear bomb detonated in 1945. Watching the fireball of the explosion from 16 km away, Fermi dropped some small pieces of paper. He watched how they fell before, during and after the explosion. He estimated that the strength of the bomb was approximately 10 kilotons. The actual value was 20 kilotons.

Worked example

A drop of vegetable oil, 0.5 mm in diameter, is dropped onto the surface of a tray of water. The oil spreads out across the water in a circular shaped layer with a diameter of 250 mm. Assuming that the layer is one molecule thick, estimate the size of a vegetable oil molecule.

Answer

The volume of the drop of vegetable oil remains the same whether it is as a drop or as the circular layer.

Drop volume — approximate the drop to a cube:

$$\text{volume} = (0.5 \times 10^{-3}\,\text{m})^3 \approx 1 \times 10^{-10}\,\text{m}^3$$

Layer volume — approximate the layer of oil on water to a square:

$$\text{volume} = (0.25\,\text{m})^2 \times \text{thickness} \approx (1 \times 10^{-1})\,\text{m}^2 \times d$$

where d is the thickness of one molecule.

Equating the two volumes gives:

$$1 \times 10^{-10}\,\text{m}^3 = 1 \times 10^{-1} \times d$$

So $d \approx 1 \times 10^{-9}\,\text{m}$.

This is an estimate of the largest size that a molecule may be, because the film may be more than one molecule thick.

At A-level, exponentials and logarithms are mainly found in the topics of capacitance and radioactivity. Exponential growth occurs when the variable you are measuring increases by the same proportion in each equal interval of time. Exponential decay occurs when the measured variable decreases by the same proportion in each equal interval of time.

Worked example

(A-level only) The potential difference (pd) V (in V) across a capacitor of capacitance C (in F) that is discharging through a resistor of resistance R (in Ω) decreases exponentially with time t (in s) according to the equation:

$$V = V_0 e^{-t/RC}$$

where V_0 is the initial pd at t_0.

A student is using data logging equipment to investigate the discharge of a capacitor. The capacitor has a labelled capacitance of 220 μF and is charged to 8.0 V. The capacitor is discharged through a 470 kΩ resistor. Calculate the pd across the capacitor after 60 s. →

Knowledge check 6

A student is determining the refractive index of a transparent block. Using a protractor, the student measures the angle of incidence as 35.5° and the angle of refraction as 26.0°. Calculate the refractive index of the block.

Knowledge check 7

What are the values of $\sin\theta$, $\cos\theta$ and $\tan\theta$ for the following angles:
a $\theta = 0°$, **b** $\theta = 45°$,
c $\theta = \pi$ rad, **d** $\theta = \pi/2$ rad.

Exam tip

If a question involves angles in radians, make sure that you use your calculator in the radian or rad mode for your calculations.

Answer

Using:

$$V = V_0 e^{-t/RC} \text{ where } V_0 = 8.0\,\text{V gives}$$

$$V = 8.0\,\text{V} \times e^{-\frac{60\,\text{s}}{470\times10^3\,\Omega\times220\times10^{-6}\,\text{F}}} = 8.0\,\text{V} \times e^{-0.580}$$

$$V = 8.0\,\text{V} \times 0.560 = 4.5\,\text{V}$$

Handling data

During most practical work you will collect data of some form. You will need to process the data in a suitable way to allow you to draw conclusions about the experiment.

Throughout your studies you will need to:

- use an appropriate number of significant figures (sf)
- find arithmetic means
- understand simple probability
- make order-of-magnitude calculations
- identify uncertainties in measurements and use simple techniques to determine uncertainty when data are combined by addition, subtraction, multiplication, division and raising to powers

Significant figures are the figures in a number that are meaningful or useful. When doing a practical, the number of significant figures recorded in your data will depend on the precision of the equipment you are using. You can find more detail about precision and uncertainties on pages 21–26.

Using experimental data to calculate other variables with your calculator will often result in an answer with a lot of decimal places. When writing out the answers, we round up or down to give the answer to the same number of significant figures as the original data.

Worked example

A student is measuring the Young modulus of a copper wire. She takes the following measurements of the diameter of the wire at five different places along the wire using a micrometer:

0.37 mm	0.38 mm	0.36 mm	0.38 mm	0.37 mm

Calculate:

a the mean diameter of the wire

b the cross-sectional area of the wire

Give your answers to an appropriate number of significant figures. →

Knowledge check 8

Do the following calculations, giving your answers in standard form to the appropriate number of significant figures.

a $\dfrac{4}{3}\pi\left(6.37\times10^6\right)^3$

b $430\,\text{THz} \times 700\,\text{nm}$

c $(0.50 \times 6.0) +$ $(\frac{1}{2} \times 9.8 \times (6.0)^2)$

Exam tip

How many significant figures? When carrying out a calculation using measured data with different numbers of significant figures, your result should contain the same number of significant figures as the measurement with the *smallest* number of significant figures (unless a detailed analysis of the uncertainties suggests otherwise).

Answer

a mean diameter $= \dfrac{(0.37 + 0.38 + 0.36 + 0.38 + 0.37)\,\text{mm}}{5}$

mean diameter $= 0.372\,\text{mm} = 3.72 \times 10^{-4}\,\text{m}$

However, while our calculator gives us 3 sf, our data have only 2 sf, so the answer should be given as

mean diameter $= 3.7 \times 10^{-4}\,\text{m}$

b cross-sectional area $= \pi r^2 = \pi(\tfrac{1}{2} \times 3.72 \times 10^{-4}\,\text{m})^2 = 1.08687 \times 10^{-7}\,\text{m}^2$

Again, rounding down to 2 sf, the answer should be:

cross-sectional area $= 1.1 \times 10^{-7}\,\text{m}^2$

Exam tip

You should use more significant figures in follow-on calculations than are needed in your final answer. Rounding before you have your final answer can introduce rounding errors and mean that your answer is wrong.

Don't forget to halve the diameter to find the radius.

Worked example

A student measures the potential difference across a resistor that is being used as part of a potential divider circuit. The potential difference is measured as 8.0 V. The current in the resistor is 2.96 mA. What is the resistance of the resistor?

Answer

$$R = \frac{V}{I} = \frac{8.0\,\text{V}}{2.96 \times 10^{-3}\,\text{A}} = 2702.7\,\Omega$$

The data in the question are given to 2 sf and 3 sf. The answer should therefore be given to just 2 sf. However, that would be $2700\,\Omega$, which could be taken to mean 2, 3 or 4 sf. In this case, it is more useful to give the resistance as $2.7\,\text{k}\Omega$.

Probability is a measure of the likelihood of an event occurring. In physics, you will mainly meet probability in the topic of radioactive decay. The probability of a nucleus decaying in a unit time is constant, known as the decay constant, λ. This gives us the equation for the *rate* of decay (or activity A) as $\dfrac{dN}{dt} = -\lambda N$, the minus sign indicating that the activity *decreases* with time.

Knowledge check 9

Without using a calculator, find the order of magnitude of the following:
a $10^{15} \times 10^3$
b $10^9 \div 10^6$
c $10^5 \times 10^{-3}$
d $10^6 \div 10^{-3}$

Algebra

Algebraic equations are used to clearly express the relationship between variables in a physical situation. To solve many problems given in A-level physics, you will need to be able to manipulate equations.

Throughout your studies you will need to:
- understand and use the symbols =, <, <<, >>, >, ∝, ≈, Δ
- change the subject of an equation, including non-linear equations
- substitute numerical values into algebraic equations using appropriate units for physical quantities
- solve algebraic equations, including quadratic equations
- **use logarithms in relation to quantities that range over several orders of magnitude**

Knowledge check 10

Estimate the following to the nearest order of magnitude.
a The number of pieces of popcorn it would take to fill a room.
b The number of electrons you have in your body.
c The number of chocolate bars you would need to eat to have sufficient energy stores to climb Mount Everest.

Symbols are a useful form of shorthand in physics. You will be familiar with many of these symbols from earlier in your school career. Some are shown in Table 3.

Table 3 Some symbols you should know

Symbol	Meaning
<<	much less than
>>	much greater than
∝	proportional to
≈	approximately equal to
Δ	change in

Worked example

(**A-level only**) A student is measuring the specific heat capacity of a block of aluminium. She uses a small heater to heat the block and a thermometer inserted into the block to measure the temperature rise. She obtains the following values:

Mass of block/kg	Energy supplied/J	Initial temperature/°C	Final temperature/°C
0.996	2.6×10^4	18	48

Calculate the value of specific heat capacity for the block of aluminium.

Answer

energy supplied = mass × specific heat capacity × temperature change

$$E = mc\Delta\theta$$

First divide both sides by $m\Delta\theta$ to get:

$$\frac{E}{m\Delta\theta} = \frac{mc\Delta\theta}{m\Delta\theta}$$

Cancelling $m\Delta\theta$ from the right-hand side of the equation we get:

$$\frac{E}{m\Delta\theta} = c$$

We can now substitute the values from the experiment into the equation:

$$c = \frac{2.6 \times 10^4\,\text{J}}{0.966\,\text{kg} \times 30\,\text{K}} = 897.17\,\text{J}\,\text{kg}^{-1}\text{K}^{-1}$$

The smallest number of significant figures in the data recorded by the student is two. The answer should therefore be given to 2 sf:

$$c = 9.0 \times 10^2\,\text{J}\,\text{kg}^{-1}\,\text{K}^{-1}$$

Knowledge check 11

Write the following statements as equations using the appropriate mathematical symbols.

a The work done in an electrical circuit is equal to the potential difference multiplied by the charge.

b The extension of a spring is directly proportional to the applied force on the spring.

c **Magnetic flux is equal to the magnetic flux density multiplied by area normal to the magnetic field**

Exam tip

You are less likely to make a mistake in calculations if you rearrange the appropriate equations *before* you put the numbers in.

Exam tip

Remember that a temperature *difference* in °C is the same as a temperature *difference* in K. For example, here we have $(48 - 18)°\text{C} \equiv 30\,\text{K}$.

Knowledge check 12

Rearrange the following equations so that the quantity shown in square brackets after each equation is the subject of the rearranged equation.

a $\rho = \dfrac{m}{V}$ [V]

b $p = mv$ [m]

c $V_{out} = V_{in} \times \dfrac{R_2}{R_1 + R_2}$ [R_2]

d $v^2 = u^2 + 2as$ [a]

e $n\lambda = d\sin\theta$ [θ]

f $F = m\omega^2 r$ [ω]

Quadratic equations contain squared terms. The most common quadratic equation that you will meet in A-level physics is the equation of motion:

$$s = ut + \frac{1}{2}at^2$$

Logarithmic scales are used to express data that have a range of several orders of magnitude. Sound is one common phenomenon that is measured using a logarithmic scale. Sound intensity is measured in $W\,m^{-2}$. A typical person can hear very quiet (low-intensity) sounds and extremely loud (high-intensity) sounds. To make a noise *sound* twice as loud, the intensity has to be approximately 10× greater. It would be very difficult to use a linear scale to represent this. A special logarithmic scale, known as the decibel (dB) scale, is used (Table 4). On this scale, a doubling of intensity is given by an increase of 10 dB.

Table 4 Sound intensity on dB scale

Sound	Intensity/$W\,m^{-2}$	Intensity/dB
Whisper	1×10^{-10}	20
Bird call	1×10^{-8}	40
Normal speech	1×10^{-6}	60
Traffic noise	1×10^{-5}	70
Pop/rock concert	1×10^{-1}	110
Pain threshold	1×10^{1}	130

Graphs

It has been said that a picture is worth a thousand words. This is definitely true in physics, where graphs are used to display data, to investigate relationships between variables and to calculate the magnitude of variables.

Throughout your studies you will need to:

- translate information between graphical, numerical and algebraic forms
- plot two variables from experimental or other data
- understand that $y = mx + c$ represents a linear relationship
- determine the slope and intercept of a linear graph
- calculate rate of change from a graph showing a linear relationship
- draw and use the slope of a tangent to a curve as a measure of rate of change
- distinguish between instantaneous rate of change and average rate of change

Knowledge check 13

A ball is dropped from a height of 1.5 m. How long does it take the ball to reach the ground?

Knowledge check 14

a Use the equation
  moment = Fd
  to calculate the moment when $F = 225\,N$ and $d = 7\,cm$.

b Use the equation
  $\varepsilon = \dfrac{x}{L}$
  to calculate L when $\varepsilon = 0.1$ and $x = 10\,cm$.

c **Use the equation**
  $x = A\cos\omega t$
  to calculate x when
  $A = 0.040\,m$,
  $\omega = \pi/3\,rad\,s^{-1}$ and
  $t = 11\,s$.

Knowledge check 15

Suggest another physical phenomenon that is measured using a logarithmic scale.

- understand the possible physical significance of the area between a curve and the x-axis and be able to calculate it or estimate it by graphical methods as appropriate
- apply the concepts underlying calculus to solve equations involving rates of change using a graphical method or spreadsheet modelling
- **interpret logarithmic plots**
- **use logarithmic plots to test exponential and power-law variations**
- sketch physical relationships that are modelled by:
  - $y = k/x$
  - $y = kx^2$
  - $y = kx$
  - $y = \sin x$
  - $y = \cos x$
  - $y = e^{\pm x}$
  - $y = \sin 2x$
  - $y = \cos 2x$

In general, when plotting graphs, the independent variable is plotted on the x-axis, and the dependent variable is plotted on the y-axis. You can read more about the practical skills involved in using graphs on pages 42–51.

A straight-line graph has the form $y = mx + c$, where m is the gradient of the graph and c is a constant. In practical work, the aim is to rearrange the equation describing the relationship between the variables into this form. If the relationship is correct, then your plot will be a straight line. Plotting data in this form also allows anomalous points to be seen more easily.

Exam tip

Current–voltage characteristics and Hooke's law graphs are often plotted with the dependent variable on the x-axis. This is convenient as it makes the gradient equal to the constant of proportionality in each case (resistance and spring constant, respectively). Look carefully at the axes on graphs in case this has been done.

Worked example

A student is using a model of a rollercoaster track (Figure 1) to investigate the relationship between the initial height of a ball on the track, and its speed at the top of the loop.

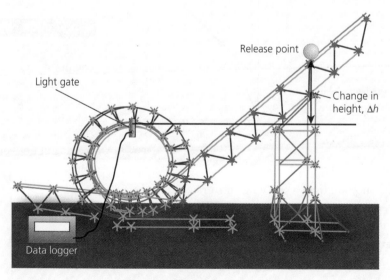

Figure 1

She varies the initial height of the ball and measures the speed at the top of the loop using a light gate and data logging software. The initial energy of the ball is equal to its gravitational potential energy $E_p = mgh$. At the top of the loop its kinetic energy is $E_k = \frac{1}{2}mv^2$.

a What quantities should the student plot in order to obtain a straight-line graph?

b What does the gradient of the graph represent?

Answer

a In this case, $\Delta E_k = \Delta E_p$ so she can write:

$$mg\Delta h = \frac{1}{2}mv^2$$

Rearranging to get an equation in the form $y = mx + c$:

$$v^2 = 2g\Delta h$$

The student will need to plot v^2 (on the y-axis) against Δh (on the x-axis) to obtain a straight-line graph.

b The gradient is $m = 2g$, where g is the acceleration of free fall.

The gradient of any graph represents the change in y for a corresponding change in x:

$$\text{gradient} = \frac{\Delta y}{\Delta x}$$

This is a *rate of change*, and forms the basis of calculus.

To calculate the gradient from a straight-line graph, we draw a large triangle using the line of best fit as the hypotenuse of the triangle, as shown in Figure 2. In this case, the gradient is equal to the rate of change of velocity with time $\left(\dfrac{\Delta v}{\Delta t}\right)$, or acceleration.

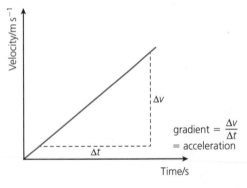

Figure 2

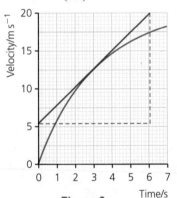

Figure 3

To calculate the gradient at a particular point from a curved graph, draw a tangent to the curve at that point. Then draw a large triangle using the tangent as the hypotenuse, as shown in Figure 3, and use this to calculate the gradient.

Mathematically, the gradient at a point is given by $\dfrac{dy}{dx}$ and can be found by differentiation.

Figures 2 and 3 both show the motion of an object that is accelerating, but in Figure 3 the acceleration is changing. The gradient at a point represents the instantaneous acceleration of the object at that point. To calculate the average acceleration of the object, the readings over the whole of the graph need to be used.

As well as the gradient of a graph enabling us to determine a rate of change, the area under a graph may also have physical significance. In velocity–time graphs, such as Figures 2 and 3, the area under the graph represents the distance travelled. In order to calculate the appropriate area under the graph, your plot must include the origin.

(A level only) Sometimes, it is helpful to plot data using logarithmic plots because the data cover a wide range of values. One way to do this is to use special log graph paper. Figure 5 shows the radioactive decay of strontium with time plotted on log-linear paper. Although the decay is exponential, the graph is linear because it has been plotted on a logarithmic scale.

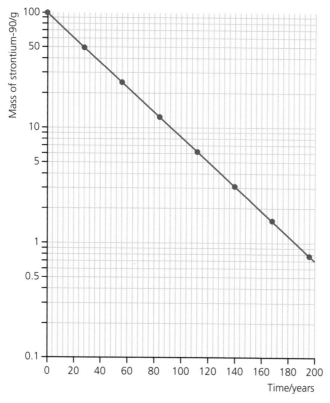

Figure 5

Instead of using special graph paper, a logarithmic plot of an exponential can also be drawn by taking natural logarithms ('ln' on your calculator) of both sides of the equation that describes the physical phenomenon.

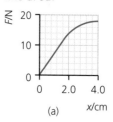
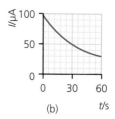

Content Guidance

Worked example

(**A level only**) A student measures the current, I, in a circuit as a capacitor discharges over time, t.

a What does she have to plot to obtain a straight-line graph?

b What do the intercept and gradient of the graph represent?

Answer

a The equation that describes the current is:

$$I = I_0 e^{-t/RC}$$

Taking natural logarithms ('ln') of both sides gives:

$$\ln I = \ln I_0 - \frac{t}{RC}$$

Plotting $\ln I$ against t will give a straight-line graph, as R and C are constants.

b The gradient is equal to $-\dfrac{1}{RC}$.

The intercept is equal to $\ln I_0$, where I_0 is the current at $t = 0$.

> **Exam tip**
>
> The graph of any exponential function of the form $y = a\,e^{bx}$ will give a straight line when we plot $\ln y$ against x. The graph of a power function $y = ax^b$ will give a straight line when we plot a graph of $\ln y$ against $\ln x$, and the gradient of the line will be equal to b.

Figure 6 gives examples of the shapes of graphs for a range of different functions.

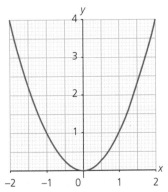

(a) Graph of $y = kx$, where k is a constant

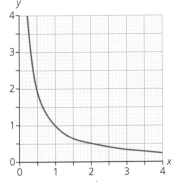

(b) Graph of $y = \dfrac{k}{x}$, where k is a constant

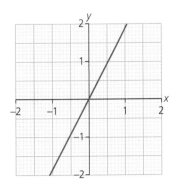

(c) Graph of $y = kx^2$, where k is a constant, in this case $k = 1$

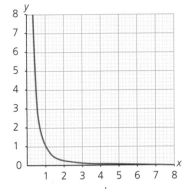

(d) Graph of $y = \dfrac{k}{x^2}$, where k is a constant, in this case $k = 1$

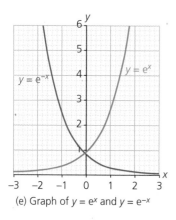

(e) Graph of $y = e^x$ and $y = e^{-x}$

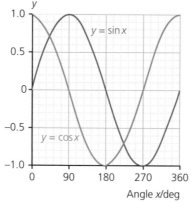

(f) Graph of $y = \sin x$ and $y = \cos x$

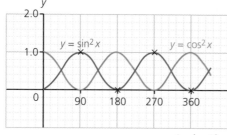

(g) Graph of $y = \sin^2 x$ and $y = \cos^2 x$

Figure 6

Geometry and trigonometry

Modelling the real world in physics requires us to solve problems in two (2D) or three (3D) dimensions. For example, we may need to think about forces acting at angles to one another; or the surface area of a solar panel normal to the sunlight.

Throughout your studies you will need to:

- use angles in regular 2D and 3D structures
- visualise and represent 2D and 3D forms, including 2D representations of 3D objects
- calculate areas of triangles, circumferences and areas of circles, and surface areas and volumes of rectangular blocks, cylinders and spheres
- use Pythagoras' theorem, and the angle sum of a triangle
- use sin, cos and tan in physical problems
- use small-angle approximations, including:
  - **$\sin \theta \approx \theta$**
  - **$\tan \theta \approx \theta$**
  - **$\cos \theta \approx 1$ for small values of θ**
- understand the relationship between degrees and radians, and translate from one to the other

When doing practical work, or analysing data from practical work, you will often have to calculate different geometric properties of objects. Table 5 gives some useful formulae for this.

Knowledge check 19

For each of the following relationships, state which sketch graph from Figure 6 they would produce.

a Q against V for $Q = CV$

b E_k against v for $E_k = \frac{1}{2}mv^2$

c g against r for $g = \dfrac{GM}{r^2}$

Content Guidance

Table 5 Some useful formulae for geometric properties

Practical activity	Shape	Equation
Area under an F–x graph for a spring obeying Hooke's law	Triangle	area = $\frac{1}{2}$ × base × height
Cross-sectional area of a wire	Circle of radius r	area = πr^2
Calculating energy radiated by an incandescent bulb	Sphere of radius r	surface area = $4\pi r^2$
Calculating energy radiated by a fluorescent tube	Cylinder of radius r and height h	surface area = $2\pi r^2 + 2\pi rh$
Calculating density of a block	Rectangular block of sides a, b and c	volume = $a \times b \times c$
Calculating density of a marble	Sphere of radius r	volume = $\frac{4}{3}\pi r^3$
Calculating the volume of a gas cylinder	Cylinder of radius r and height h	volume = $\pi r^2 h$

> **Exam tip**
>
> Data are often given for the diameter of a circle or sphere, instead of the radius. Remember that radius = ½ × diameter and so $r^2 = \frac{1}{4}d^2$.

When calculating the resultant of forces that are applied to an object, knowledge of trigonometry is essential.

Pythagoras' theorem states that, if a right-angled triangle has sides a, b and c, where c is the hypotenuse, then $a^2 + b^2 = c^2$ (Figure 7). Right-angled triangles are also used to define the three functions, sine, cosine and tangent.

- $\sin\theta$ = opposite/hypotenuse
- $\cos\theta$ = adjacent/hypotenuse
- $\tan\theta$ = opposite/adjacent

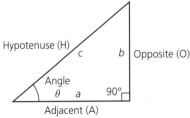

Figure 7

> **Knowledge check 20**
>
> A copper pipe has an internal diameter of 12 mm and a length of 150 cm. Calculate the volume of liquid that can be contained in the pipe.

> **Knowledge check 21**
>
> The molar mass of carbon is $12.0 \times 10^{-3}\,\mathrm{kg\,mol^{-1}}$ and the density of diamond is $3500\,\mathrm{kg\,m^{-3}}$. Estimate the atomic radius of a spherical carbon atom.

> **Exam tip**
>
> The sum of the internal angles in a triangle is 180°.

> **Exam tip**
>
> 1 radian = 57.3°
>
> Angle in radians = $\dfrac{\pi}{180°}$ × angle in degrees
>
> Angle in degrees = $\dfrac{180°}{\pi}$ × angle in radians

Worked example

A student is investigating equilibrium using a force board as shown in Figure 8.

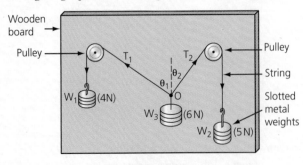

Figure 8

The system comes to rest with angles $\theta_1 = 57°$ and $\theta_2 = 42°$. Show that the system is in equilibrium.

Answer

To be in equilibrium, the forces in each direction at point O (weight 3) must balance.

Resolving horizontally: $T_1 \sin \theta_1 = T_2 \sin \theta_2$

$T_1 \sin \theta_1 = 4\,\text{N} \sin 57° = 3.4\,\text{N}$ and $T_2 \sin \theta_2 = 5\,\text{N} \sin 42° = 3.3\,\text{N}$

We can see that the forces do not balance exactly. This is probably due to friction in the system causing experimental uncertainty.

Resolving vertically: $T_1 \cos \theta_1 + T_2 \cos \theta_2 = W_3$

$4\,\text{N} \cos 57° + 5\,\text{N} \cos 42° = 5.9\,\text{N}$

Once again, the vertical components balance to within a reasonable experimental error.

Degrees and radians are used to describe angles. In many cases, the radian is a useful measure, particularly when working with waves and oscillations.

A radian is defined as the angle created by the arc s of a circle that is equal in length to the radius r of the circle ($s = r$). Figure 9 shows that an angle θ in radians is given by $\theta = s/r$ or $s = r\theta$.

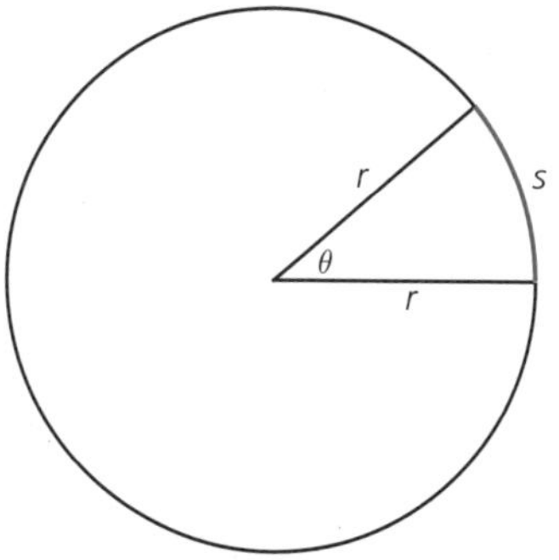

When using small angles, measured in radians, we can use the following **small-angle approximations**, as the values are very similar:

$\sin \theta \approx \theta$

$\tan \theta \approx \theta$

$\cos \theta \approx 1$

Figure 9

■ Practical skills

Physics is a practical subject, and experimental work should form a significant part of your A-level physics course. It is commonly acknowledged that it is easier to learn and remember things if you have actually done them rather than having read about them or been told about them. That is why this section of the book is illustrated throughout by experiments, usually with a set of data for you to work through and questions to answer.

As you complete practical work during the A-level physics course, you will develop skills in the use of a range of apparatus and techniques. Examination questions will test your knowledge and understanding of these techniques. You should refer to the specification, which contains a list of apparatus and techniques with which you should be familiar.

Knowledge check 22

Express the following angles in radians to 3 sf: **a** 15.0°, **b** 50.0°, **c** 145°.

Knowledge check 23

Express the following angles in degrees to 3 sf: **a** $\pi/8$ rad, **b** 1.50 rad, **c** 0.150 rad.

Knowledge check 24

Calculate $\sin \theta$, $\cos \theta$ and $\tan \theta$ to 2 sf for the following angles: **a** 5.0 rad, **b** 0.50 rad, **c** 0.050 rad.

Exam tip

Make sure you have your calculator set in the right mode — degrees or radians — as required by the question.

Practical skills assessed in written papers

Examiners may ask you questions about your knowledge of scientific procedures or they may test your scientific knowledge set in a range of practical contexts. Although the focus will be on your physics knowledge, without a good understanding of practical work you will find it difficult to answer this type of question. You will be expected to apply skills gained from the practical endorsement and will be required to describe and evaluate specific methods and procedures. These could include questions about alternative and unfamiliar methods for the experiments you may have carried out during your A-level physics course. This means that it is vital that you understand the practical aspects of your A-level physics course really well, so that you are able to plan, implement, analyse and evaluate experiments from every topic on the specification. Approximately 15% of the marks in the written papers will assess the following practical skills:

- PS1.1   Solve problems set in practical contexts
- PS1.2   Apply scientific knowledge to practical contexts
- PS2.1   Comment on experimental design and evaluate scientific methods
- PS2.2   Present data in appropriate ways
- PS2.3   Evaluate results and draw conclusions with reference to measurement uncertainties and errors
- PS2.4   Identify variables including those that must be controlled
- PS3.1   Plot and interpret graphs
- PS3.2   Process and analyse data using appropriate mathematical skills
- PS3.3   Consider margins of error, accuracy and precision of data
- PS4.1   Know and understand how to use a wide range of experimental and practical instruments, equipment and techniques

How does practical work fit into the AS and A-level exams?

The AQA specification states that *all students taking this course must carry out the required practicals. Written papers will assess knowledge and understanding of these practicals, and the skills exemplified within each practical.*

At AS, the majority of practical-based questions will be set on paper 2, with around 20 of the 70 available marks on this paper based on practical skills. At A-level, the majority of practical-based questions will be on paper 3, and will account for around 45 of the 80 available marks. *Overall, 15% of the marks for all AS and A-level physics courses will require the assessment of practical skills.*

About this section of the book

For convenience, this section of the practical guide is in five main sections, which reflect an orderly approach to practical work:
- Planning
- Making measurements
- Recording data
- Analysing results
- Evaluating results and drawing conclusions

Having said this, there will inevitably be a degree of overlap between each section.

Planning

Once the problem has been identified, you will be expected to produce a plan. Planning an experiment or investigation includes selecting appropriate equipment to make the relevant measurements. You will also need to be able to describe measurement strategies and techniques to ensure accurate results. Safety issues must also be discussed.

Types of variables

You are expected to be able to identify variables, including those that must be controlled. You should be familiar with the following terminology:

- The *independent variable* is the variable for which values are changed by the experimenter.
- The *dependent variable* is the variable for which the value is measured for each and every change in the independent variable.
- *Control variables* may, in addition to the independent variable, affect the outcome of the investigation and therefore have to be kept constant or at least monitored.

Worked example

A student investigates how the diameter of a ball bearing affects how long it takes to fall through a measuring cylinder filled with oil. What are the independent, dependent and control variables in this experiment?

Answer

Independent variable is diameter of ball bearing.

Dependent variable is time.

Control variables are temperature of oil, distance fallen, material of ball bearing, smoothness of ball bearing surface, type of oil and height of release of ball bearing.

A key part of planning an experiment is describing how to obtain accurate and precise data. This involves choosing the most appropriate equipment and using it in such a way to enable the most accurate data to be obtained. We will look at good experimental design on page 28 after the section on errors and uncertainties.

Making measurements

Errors, accuracy and precision

One of the specified requirements is that you should *consider margins of error, accuracy and precision of data*. In everyday English, accuracy and precision have similar meanings, but in physics this is not the case. Their meanings are not the same and the difference must be understood.

An error is the difference between a measured result and the true value or accepted value, i.e. the value that would have been obtained in an ideal measurement (which is impossible to achieve!). With the exception of a fundamental constant, the true value is considered unknowable. An error can be due to random or systematic effects, and an error of unknown size is a source of uncertainty.

Random errors

There are always random errors present due to the way any instrument works, the way it is used or changes in the external conditions. The same measurement will give different values each time it is measured. The effect of this type of error can be reduced by taking repeat measurements and then averaging or by drawing a graph. Random errors produce scatter of the points around the line of best fit (Figure 10).

Random errors will give rise to an *uncertainty* in the measurement you have taken. It is important to be able to estimate the uncertainty in a measurement so that its effect can be taken into consideration when drawing conclusions about experimental results.

Systematic errors

A systematic error produces a measurement that is consistently too large or too small by the same amount. This could be caused by recording the wrong unit, poor technique (e.g. systematically reading off from a scale at an angle, causing parallax error), failure to check for zero error, or incorrect calibration of the instrument.

Zero error

A zero error occurs when the measuring instrument is not set on zero accurately. The zero error should be subtracted or added to each measured reading to give the true value (Figure 11).

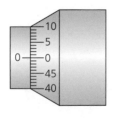 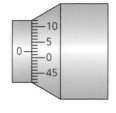 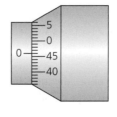

Figure 11   (a) No zero error   (b) Positive zero error   (c) Negative zero error

Sometimes a systematic error is not apparent from the measurements. Another advantage of drawing a graph is that it can indicate whether there is a systematic error and enable you to make allowance for it (Figure 12).

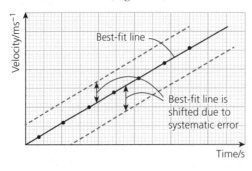

Figure 12

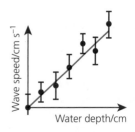

Figure 10

Exam tip

Before an instrument is used to make a measurement, it is important to check for zero error. This is an example of a systematic error, which will cause a constant error in all the readings.

Knowledge check 26

The thickness of a sheet of glass is measured to be 0.86 mm using the micrometer shown in Figure 11(b). What is the actual thickness of the glass?

A systematic error will be seen on a graph by the line of best fit being shifted up or down the *y*-axis because each measurement is larger or smaller than the true value by the same amount. A systematic error does *not* affect the *gradient* of the line of best fit but simply alters the *y*-intercept.

Accuracy and precision

A measured value is considered to be **accurate** if it is judged to be close to the true value. Any actual measurement will always be subject to random and systematic errors.

The term **precision** denotes the consistency between values obtained by repeated measurements — a measurement is precise if the values 'cluster' together. Precision is influenced only by random errors. We can try to ascertain and quantify these errors, i.e. *estimate the uncertainty* in our value. Precise results would show little scatter around the line of best fit on a graph but might have a systematic error, making them inaccurate. Accuracy and precision are summarised in Figure 13.

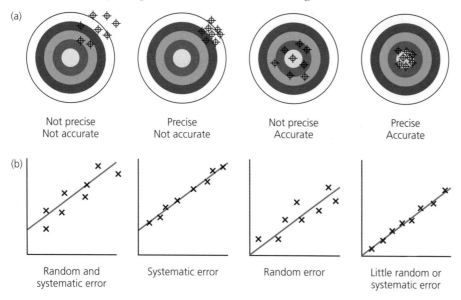

(a)

Not precise
Not accurate

Precise
Not accurate

Not precise
Accurate

Precise
Accurate

(b)

Random and
systematic error

Systematic error

Random error

Little random or
systematic error

Figure 13

Uncertainties

Every measurement has an inherent uncertainty due to the resolution of the instrument, manufacturing error, the way the instrument is used (human error!) or the design of the experiment.

An uncertainty for a single reading, or repeated readings that are the same, will be at least half the **resolution** of the instrument. For repeated readings that are different, the uncertainty can be taken as being *half the range* of the readings.

As a general guide, the uncertainty of a **reading** is at least ±0.5 of the resolution. A reading is a value found from a single judgement, e.g. thermometer, Geiger counter, top-pan balance, measuring cylinder, or ruler with a zero end. For such 'analogue' devices, it may be possible to interpolate to the nearest half a division on the scale (Figure 14).

Knowledge check 27

Explain the difference between accuracy and precision.

Knowledge check 28

The magnetic field strength of the Earth at a point on the Earth's surface was measured five times and gave the following values: 50 mT, 48 mT, 46 mT, 52 mT, 62 mT. The true value of the magnetic field strength at this point was known to be 51 mT. Comment on the precision of these data.

Figure 14

The uncertainty of a **measurement** is at least ±1 of the resolution. A measurement requires the value to be taken from two readings or judgements, e.g. non-zero-ended ruler (the start and the end readings), stopwatch (start and stop) or extension of a spring (original and final lengths).

Worked example

Estimate the uncertainty in a length measurement of 122 mm made using the ruler shown in Figure 15, and write down how the length should be recorded.

Figure 15

Answer

The smallest division on the ruler is 1 mm. Therefore, the uncertainty in each reading is ½ × 1 mm = 0.5 mm.

But as we are taking readings at each end, we are making two judgements. So the total uncertainty in this measurement of length would be ±(2 × 0.5 mm) = ±1 mm.

Length = 122 mm ± 1 mm.

Other factors affecting the uncertainty in a measurement

The general guidelines above give the *smallest* uncertainty in a measurement or reading. Often the resolution of the instrument is not the limiting factor, and the way the instrument is used can lead to larger uncertainties. The instrument's resolution allows us to assess the *minimum* possible uncertainty. You need to make an assessment of the other factors, such as the set-up and use of the apparatus, and take these into account.

In experiments, you will often be timing events or measuring the length of something. The following examples should help you understand how the uncertainties involved are actually much larger than those due to the resolution of the stopclock or ruler.

A stopwatch typically has a resolution of 0.01 s, but the reaction time of the person using it is likely to be no quicker than 0.1 s or 0.2 s. You should record the full reading on the stopwatch (e.g. 18.62 s) and reduce this to a more appropriate number of significant figures at a later stage, e.g. after averaging readings.

If you are measuring the length of a piece of an unstretched elastic band, it is difficult to place the elastic completely straight against the ruler. This could lead to an uncertainty of ±2 or ±3 mm.

The uncertainty of the reading from digital voltmeters and ammeters depends on the tolerance quoted by the manufacturer (which may be considerable). If this is not known, then the reading should usually be stated to ±1 of the last digit when quoting the value and the uncertainty.

Exam tip

When using an analogue instrument, always try to interpolate between whole divisions, e.g. to the nearest 0.5 mm on a ruler or to 0.1 degrees for a thermometer calibrated in degrees.

Exam tip

In exam papers, values will often be stated together with the absolute uncertainty, e.g. 'a wire has a diameter of 1.87 mm ± 0.01 mm'. The uncertainty should be quoted to be the same number of decimal places as the reading.

Knowledge check 29

What is meant by the resolution of an instrument?

Exam tip

Always write down the full reading and then, if necessary, round to fewer significant figures when the uncertainty has been estimated.

Knowledge check 30

A 100 Ω resistor has a tolerance of 5%. Calculate the minimum and maximum values it could have.

Worked example

Estimate the uncertainty in the reading of potential difference shown in Figure 16, and state how you would record the value of the pd. The dial is set on the 20 V dc scale.

Figure 16

Answer

The uncertainty will be ± the smallest division shown on the meter = ±0.01 V.

The reading is therefore 19.16 V ± 0.01 V.

The manufacturing tolerance is not stated, so by quoting the uncertainty to ± the smallest division, we are making a best possible estimate.

Percentage uncertainties

It is useful to be able to quote uncertainties as percentages. This also enables uncertainties to be combined easily.

$$\text{percentage uncertainty} = \frac{\text{uncertainty}}{\text{value}} \times 100\%$$

Worked example

The digital voltmeter shown in Figure 16 is being used to measure the pd across a lamp. Estimate the percentage uncertainty in this reading.

Answer

The uncertainty will be ± the smallest division shown on the meter = ±0.01 V.

The reading is therefore 19.16 V ± 0.01 V.

Estimated percentage uncertainty = (0.01 V/19.16 V) × 100% = 0.05%

Exam tip

The manufacturer's tolerance will almost certainly be much greater than 0.05%, and for many digital meters it can be as much as 5% or even 10%.

Combining uncertainties

In experiments, measurements are often made using several instruments and then used to calculate a quantity. The combined uncertainty can be found as follows:

- If quantities are added or subtracted, *add* the *absolute* uncertainties of each quantity.
- If quantities are multiplied or divided, *add* the *percentage* uncertainties of each quantity.
- If a quantity is raised to a power, then *multiply* the *percentage* uncertainty by the *power.*

Worked example

The pd across a bulb is measured to be 3.15 V when the current in it is 40.8 mA.

a Calculate the power.

b Estimate the percentage uncertainty in the power.

c State how the power should be recorded.

Answer

a Power = VI = 3.15 V × 40.8 × 10^{-3} A = 0.129 W (data to 3 sf, and so answer should be to 3 sf)

b The pd and current are both *readings*, so the uncertainty is *theoretically* half the smallest division, but in practice we take it to be the smallest division to make some allowance for manufacturing tolerance of the digital meters. In this case, 0.01 V and 0.1 mA. (We shall often write % uc as shorthand for percentage uncertainty.)

So, % uc in V = (0.01 V/3.15 V) × 100% = 0.317%

and % uc in I = (0.1 mA/40.8 mA) × 100% = 0.245%

This gives % uc in P = % uc in V + % uc in I

so % uc in P = 0.317 + 0.245 = 0.562% = 0.6%

c Uncertainty in P = 0.562% of 0.129 W = 7.2 × 10^{-4} W ≈ 0.001 W

Power = 0.129 W ± 0.001 W or 129 mW ± 1 mW

Recording data

You will be expected to *present data in appropriate ways*. You may also need to *process the data using appropriate mathematical skills.*

Tabulating data

Data must be clearly recorded in a table. Tables should have headings with the quantity and units separated by a forward slash (e.g. current/mA). You should also ensure that the number of significant figures, or decimal places, is appropriate, and these should be consistent down each column. The independent variable should normally be the first column of the table.

Table 6, containing the results from an experiment to measure electrical power, shows how data should be recorded.

Knowledge check 31

The diameter of a ball bearing is measured using vernier callipers with 0.1 mm divisions. The vernier reading is 21.5 mm.

a State the uncertainty in this measurement, and write down how the value for the diameter should be recorded.

b Estimate the percentage uncertainty in this value.

Exam tip

Keep the data in your calculator and only round the final answer.

Knowledge check 32

What is wrong with this table?

Volts	A
5.85 V	0.05
4.38 V	0.04
2.94 V	0.03
1.4 V	0.01

Table 6

Potential difference V/V	Current I/mA
6.01	58.8
4.63	50.8
3.31	41.8
2.16	32.9

Worked example

(**A-level only**) In an experiment to investigate whether radioactive decay is exponential, a student decides to plot a graph of the natural logarithm of the activity A, which is measured in counts per second, against time t, which is measured in seconds. Draw a table showing the headings that the student should use in order to record these data.

Answer

Time t/s	Activity A/s^{-1}	ln $(A$/s$^{-1})$

Significant figures

When recording data, it is important to *write down* the data to the number of significant figures given by the resolution of the device being used. For digital instruments, you should write down all of the numbers shown on the device, e.g. times should be *recorded* to 0.01 s if using a digital stopwatch. Only in the subsequent processing should you round down to the appropriate number of significant figures. The following examples show how to correctly use significant figures in recorded and processed data.

Worked example

In an experiment to find the period T of a simple pendulum, the following times are recorded for 20 oscillations:

t_1/s	t_2/s	t_3/s	Mean t/s	T/s	T^2/s^2
14.72	14.65	14.67			

Complete the table by adding values of t, T and T^2.

Answer

Mean $t = 14.68$ s $T = t/20 = 0.734$ s $T^2 = 0.539$ s^2

Worked example

The following data were recorded for a trolley accelerating down an inclined plane:
- initial velocity of the trolley = 1.23 m s^{-1}
- final velocity of the trolley = 1.97 m s^{-1}
- time taken = 1.6 s

Use these data to calculate the acceleration of the trolley, quoting your answer to the appropriate number of significant figures.

Answer

$$a = (v - u)/t = (1.97 - 1.23)\,\text{m s}^{-1}/1.6\,\text{s}$$
$$= 0.4625\,\text{m s}^{-2} = 0.46\,\text{m s}^{-2}$$

(Answer is to 2 sf because the time is only recorded to 2 sf even though the velocities are recorded to 3 sf.)

Good experimental design

When planning an experiment, you should attempt to design a method that leads to the smallest uncertainties in your measurements. Exam questions may ask you to look at a number of factors:

- the resolution of the instrument used
- the manufacturer's tolerance on instruments and components such as resistors and capacitors
- how the experimenter uses the instruments (e.g. parallax, zero error)
- the procedures adopted (e.g. repeated readings)

You may be required to evaluate procedures and explain how particular techniques could affect uncertainties in the measurements, or how the uncertainties could be reduced by using different apparatus or procedures.

Parallax error

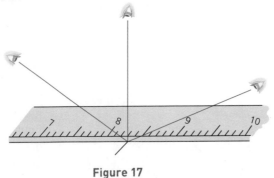

Figure 17

When reading a scale, for example from an analogue meter or ruler, you should ensure that the pointer, scale and your eye are all in line. This will avoid *parallax error*. It is important not look at the scale from an angle, as this would give a measurement that is larger or smaller than the true value (Figure 17).

Using a vernier scale

Vernier callipers and micrometers enable measurements of length or thickness to be made to a much greater resolution than using a ruler and can reduce the uncertainty by a factor of 10 or 100, respectively. Often you will be able to use electronic or digital instruments, which have an even greater resolution.

Knowledge check 33

In an experiment to measure the acceleration of a trolley, the acceleration was measured three times using two light gates and a data logger:

Acceleration/m s^{-2}		
0.44	0.45	0.47

Calculate the mean acceleration, quoting your answer to the appropriate number of significant figures.

Knowledge check 34

a Figure 17 shows an enlarged millimetre scale. What is the correct reading?
b Explain how a mirror can be used to reduce parallax error when taking a reading from an analogue scale.

Vernier callipers measure to the nearest 0.1 mm and therefore measure to a precision of 1% or better if used to measure lengths of 1 cm or longer.

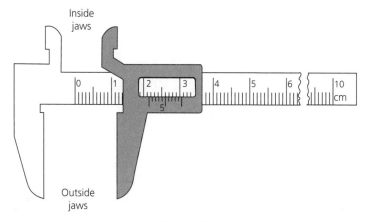

Inside jaws

Outside jaws

Figure 18

Figure 18 shows how to read the scale — a hand lens is recommended, as the scale lines can be difficult to see:

- the zero on the vernier scale is just beyond the 2.1 cm (21 mm) mark on the main scale
- next locate the lines on the main scale and the vernier scale that are *exactly lined-up with each other*
- in this case this is the '5' on the vernier scale
- the reading is therefore 2.15 cm (21.5 mm)

Vernier scales are used in a number of instruments, for example travelling microscopes, but the principle of reading the scale is always the same.

The following example shows how, by using a vernier scale and a sensitive electronic balance, the accuracy of the measurement of length and mass, and therefore the determination of density, can be improved. This is particularly important when a quantity is being squared or cubed.

Worked example

a   A student is asked to determine the density of aluminium in the form of a cylinder. He measures the diameter and length of the cylinder with a metre rule and the mass with a top pan balance. He records the following measurements:

length = 75 ± 1 mm

diameter = 25 ± 1 mm

mass = 98.8 ± 0.1 g

Use these data to determine the student's value for the density of aluminium, and estimate the percentage uncertainty in his value.

b   His teacher suggests that he would get a more accurate value for the density if he used vernier callipers to measure the diameter and length.

What does the teacher mean by 'accurate'. Estimate the improvement to the accuracy this would make.

→

Knowledge check 35

What does the vernier scale shown in Figure 19 read?

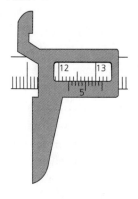

Figure 19

Knowledge check 36

Calculate the density of a marble having a diameter of 10 mm and a mass of 1.3 g. Give your answer in $kg\,m^{-3}$ and in $g\,cm^{-3}$.

Exam tip

You are less likely to make a mistake if you let the calculator do the work when converting units rather than doing it in your head — put in the radius of 12.5 mm as $12.5 \times 10^{-3}\,m$ rather than trying to work out 0.0125 m.

Answer

a volume = $\pi r^2 h = \pi \times (12.5 \times 10^{-3}\,\text{m})^2 \times 75 \times 10^{-3}\,\text{m} = 3.68 \times 10^{-5}\,\text{m}^3$

  density = mass/volume = $98.8 \times 10^{-3}\,\text{kg}/3.68 \times 10^{-5}\,\text{m}^3$

  $= 2.7 \times 10^3\,\text{kg}\,\text{m}^{-3}$ (2 sf in line with the data)

  % uc in $d = (1\,\text{mm}/25\,\text{mm}) \times 100\% = 4.0\%$

  so % uc in $r^2 = $ % uc in $d^2 = 2 \times 4.0\% = 8.0\%$

  % uc in $h = (1\,\text{mm}/75\,\text{mm}) \times 100\% = 1.3\%$

  % uc in $m = (0.1\,\text{g}/98.8\,\text{g}) \times 100\% = 0.1\%$

  % uc in density = $8.0\% + 1.3\% + 0.1\% = 9.4\% \approx 9\%$

b An accurate value is one that is close to the true value.

  A vernier scale will have a resolution of 0.1 mm, compared with 1 mm for the rule. The percentage uncertainties in d and h will therefore be 0.40% and 0.13%, respectively.

  The percentage uncertainty in the density will then be $(2 \times 0.40 + 0.13 + 0.10) = 1.03\% \approx 1\%$.

  This means the accuracy will be improved by a factor of about nine times.

Exam tip

When a quantity is raised to the power of n, the *percentage* uncertainty in the quantity is multiplied by n.

Knowledge check 37

a What is the resolution of the main scale of the micrometer shown in Figure 20?

b What is the thickness of the object being measured by the micrometer shown in Figure 20? What assumption are you making?

Using a micrometer

A micrometer (Figure 20) has a resolution of ±0.01 mm. The micrometer should be closed using the ratchet until it *just* clicks.

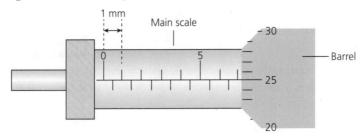

1 mm · Main scale · 30 · Barrel · 0 · 5 · 25 · 20

Figure 20

When using a micrometer, remember the following:

- Do not over-tighten, as this could squash the object and/or damage the thread. Use the *ratchet* to close the jaws.
- Check for a zero error and add or subtract as necessary.
- Make sure you read the scale correctly (in particular, whether the main scale reading is less than or more than 0.5 mm).
- Measure the diameter several times and in several places to check for uniformity. (For a wire, for example, you could measure the diameter at each end and in the middle to check if the wire is thinner at one end. You could also then take readings at right angles to check that the wire has a circular cross-sectional area.)

Exam tip

You should remember to state the precautions that should be taken to improve accuracy when using a micrometer.

Knowledge check 38

A micrometer has a zero error of +0.03 mm. A student measures the diameter of a pencil lead to be 2.12 mm. What is the actual diameter of the lead?

Repeated measurements

Random errors are generally caused either by human misjudgements, e.g. reaction time, interpolating between scale divisions, or by variations in a measurement, e.g. non-uniformity of a wire. Random errors will cause measurements to spread about a mean value. As precise measurements are measurements in which there is very little spread about the mean value, you must try to reduce random errors in order to improve precision.

Because they are likely to cause readings to be larger than the true value as often as they are smaller than the true value, repeating readings or measurements will smooth out the effect of random errors.

For repeat measurements:

$$\text{percentage uncertainty} = \frac{\frac{1}{2} \times \text{range}}{\text{mean value}} \times 100\%$$

By plotting a graph, the line of best fit will effectively average out the random errors, but they can never be eliminated completely. You will also be able to see the scatter of the points about the line of best fit and so see how precise the measurements are.

Worked example

The resistance of a thermistor was measured three times at 20°C: 670 Ω, 661 Ω, 678 Ω. Calculate the percentage uncertainty in the value of the resistance, and state how the resistance of the thermistor should be recorded.

Answer

range = 678 − 661 = 17 Ω

mean = (670 + 661 + 678)/3 = 669.7 Ω = 670 Ω (3 sf)

$$\%uc = \frac{\left(\frac{1}{2} \times 17\right)\Omega}{670~\Omega} \times 100\% = 1.27\%$$

1.27% of 669.7 Ω = 8.5 Ω ⇒ R = 670 Ω ± 9 Ω

Worked example

For an angle of incidence of 40.0° on a glass block, the angle of refraction was measured three times as: 25.5°, 25.0°, 26.0°. How should the mean value of the angle of refraction be expressed?

Answer

range = 26.0° − 25.0° = 1.0°

mean = 76.5/3 = 25.5°

angle of refraction = 25.5° ± 0.5°

Exam tip

Improving precision: you need to be able to describe how the measurements obtained from an experiment can be made more precise.

Worked example

Measuring the density of glass and sand

We can illustrate a number of the points we have made concerning good experimental design by considering an experiment to investigate whether glass might be manufactured from sand. The following apparatus is used: glass block, plastic cup, dry sand, 250 ml measuring cylinder, vernier callipers and access to balance and water.

The experiment is carried out as follows:

For the glass block:

■ Using the vernier callipers, the dimensions are determined as:

$l = (116.0 \pm 0.2)\,\text{mm}$

$w = (63.0 \pm 0.2\ )\,\text{mm}$

$h = (48.7 \pm 0.2)\,\text{mm}$

■ Using the electronic balance:

mass $= 882.4 \pm 0.1\,\text{g}$

For the sand:

■ Using the electronic balance:

mass of empty plastic cup $= 35.5 \pm 0.1\,\text{g}$

mass of cup and sand $= 87.1 \pm 0.1\,\text{g}$

■ The measuring cylinder is filled with $100\,\text{cm}^3$ of water and then the sand is added.

volume of the water + sand $= 123\,\text{cm}^3$

a Determine the densities of glass and sand.
b Estimate the uncertainties in each of these values.
c Make a comparison between your final values with reference to these uncertainties.
d Write a conclusion discussing the extent to which your results support the fact that glass is made from sand.
e Explain any experimental techniques that you would use in the above experiment to try to reduce the uncertainty in the measurements.

Answer

a *Glass:*

volume $= 116.0 \times 10^{-3}\,\text{m} \times 63.0 \times 10^{-3}\,\text{m} \times 48.7 \times 10^{-3}\,\text{m}$

$= 3.56 \times 10^{-4}\,\text{m}^3$

density = mass/volume $= 882.4 \times 10^{-3}\,\text{kg}/3.56 \times 10^{-4}\,\text{m}^3$

$= 2.48 \times 10^3\,\text{kg}\,\text{m}^{-3}$ or $2.48\,\text{g}\,\text{cm}^{-3}$

→

Sand:

volume of sand = $(123 - 100)\,cm^3 = 23\,cm^3 \pm 2\,cm^3$ (as we have two readings)

mass of sand = $(87.1 - 35.5)\,g = 51.6\,g \pm 0.2\,g$

density = mass/volume = $51.6\,g/23\,cm^3$

$\qquad = 2.2\,g\,cm^{-3}$ (to 2 sf, as volume is only to 2 sf)

b *Glass:*

% uc in mass = $(0.1\,g/882.4\,g) \times 100\% = 0.01\%$

% uc in volume = % uc in length + % uc in width + % uc in depth

$\qquad = (0.17 + 0.32 + 0.41)\% = 0.90\%$

% uc in density = % uc in mass + % uc in volume

$\qquad = (0.01 + 0.90)\% = 0.91\%$

Uncertainty in density = 0.91% of $2.48\,g\,cm^{-3} = \pm0.02\,g\,cm^{-3}$

Sand:

% uc in density = % uc in mass + % uc in volume

$\qquad = (0.2\,g/51.6\,g) \times 100\% + (2\,cm^3/23\,cm^3) \times 100\%$

$\qquad = (0.4 + 8.7)\% = 9.1\%$

Uncertainty in density = 9.1% of $2.2\,g\,cm^{-3} = \pm0.2\,g\,cm^{-3}$

c The percentage difference between the density of sand and the density of glass is:

$$\frac{(2.48 - 2.2)\,g\,cm^{-3}}{2.34\,g\,cm^{-3}} \times 100\% = 12\% \text{ smaller than the density of glass}$$

d The minimum value that the glass could have is $2.46\,g\,cm^{-3}$ and the maximum possible value for the sand is $2.4\,g\,cm^{-3}$.

As the values do not quite overlap, the 12% difference cannot be accounted for by the experimental uncertainties, suggesting that glass may *not* be made entirely from sand. More likely reasons for the discrepancy are that the sand may not have been completely dry, or not all the air between the sand particles was replaced by water.

e You might suggest the following:

- Check vernier for zero error.
- Repeat measurements of block dimensions in at least two different places for each dimension.
- Shake the sand well to prevent air pockets.
- Keep eye level with water meniscus to prevent parallax error.

Knowledge check 40

Explain how you would determine, as accurately as possible, the volume of a glass marble.

Exam tip

The percentage difference between two experimental values is given by:

$$\%\text{difference} = \frac{\text{difference between the values}}{\text{average of the two values}} \times 100\%$$

The percentage difference between an experimental value and a stated or known value is given by:

$$\%\text{difference} = \frac{\text{difference between the values}}{\text{stated value}} \times 100\%$$

Multiple readings

The use of multiple readings reduces the uncertainty. For example, if you determine the thickness of one sheet of paper by measuring the thickness of 50 sheets in a stack, and dividing by 50, you will then reduce the uncertainty in the measurement of one sheet by a factor of 50. In general, the uncertainty of each measurement will be the uncertainty of the whole measurement divided by the number of instances (e.g. sheets of paper or oscillations of a pendulum). This method works because the *percentage* uncertainty in the value of the thickness of one sheet will be the same as the percentage uncertainty in the measurement of multiple sheets.

Worked example

The thickness of ten sheets of paper is measured with a micrometer (resolution ±0.01 mm) and found to be 1.08 mm. What is the percentage uncertainty in the mean thickness of one sheet?

Compare this to a measurement of the thickness of a ream of 500 sheets made using a vernier (resolution ±0.1 mm), which is 55.1 mm.

Answer

Thickness of ten sheets of paper: 1.08 mm ± 0.01 mm = 1.08 mm ± 0.9%

Mean thickness of one sheet of paper = 1.08 mm/10 = 0.108 mm ± 0.9%

Thickness of ream of 500 sheets = 55.1 mm ± 0.1 mm = 55.1 mm ± 0.18%

Mean thickness of one sheet = 55.1 mm/500 = 0.110 mm ± 0.18%

By measuring 500 sheets, the percentage uncertainty in the mean thickness of one sheet is reduced by a factor of 5.

This technique is particularly useful when applied to timing. By timing over multiple oscillations, the overall uncertainty will be reduced significantly and the effect of the reaction time will become less significant.

Exam tip

The absolute uncertainty in a measurement made with a metre ruler is about 1 mm. Provided lengths in excess of 100 mm are measured, then the percentage uncertainty will be less than 1%.

Exam tip

Always record the actual instrument reading and round down later.

Knowledge check 41

a Why is the resolution of a stopclock (0.01 s) not used when determining the uncertainty of timing an event taking about 3 s?

b Suggest a suitable uncertainty for a single such event.

Worked example

Figure 21 shows a stopwatch reading for 20 oscillations of a simple pendulum.

a What is the resolution of the instrument?

b Discuss what a realistic estimate of the uncertainty in this time would be.

Figure 21

c Explain why it is an advantage to time 20 oscillations when determining the period T of the oscillations.

d Determine a value for the period T and the estimated uncertainty in your value.

e Compare this uncertainty with that which would be incurred if only one oscillation had been timed.

Answer

a The resolution of the stopwatch is ± 0.01 s.

b This is insignificant compared to human reaction time, which is at least 0.05 s. As this applies both when starting and stopping the timing, it would be appropriate to use ± 0.1 s as the uncertainty in the measurement of time using a stopwatch.

c By measuring the time for 20 oscillations, and then dividing by 20, you would reduce the uncertainty by a factor of 20, as the error only occurs when the stopwatch is started and stopped.

d $20T = 12.60 \text{ s} \Rightarrow T = 0.630 \text{ s}$

Estimated uncertainty is reaction time of 0.1 s:

% uc in timing $= (0.1 \text{ s}/12.60 \text{ s}) \times 100\% = 0.8\% = \%$ uc in T

uncertainty in $T = 0.8\% \times 0.630 \text{ s} = 0.005 \text{ s} \Rightarrow T = 0.630 \pm 0.005 \text{ s}$

e If only one oscillation was timed, measured T would be 0.63 ± 0.10 s:

% uc in $T = (0.1 \text{ s}/0.63 \text{ s}) \times 100\% = 16\%$

(This is, of course, 20 times as much as the percentage uncertainty when timing 20 oscillations.)

Exam tip

Timing 20 oscillations and then dividing by 20 to find the time period reduces the uncertainty by a factor of 20.

Knowledge check 42

A mass on a spring completes ten oscillations in 18.9 s. Determine the percentage uncertainty in the time period suggested by the precision of the recorded data.

Using a set-square to reduce uncertainty

A set-square can be particularly useful to help determine precise measurements, as shown in Figure 22.

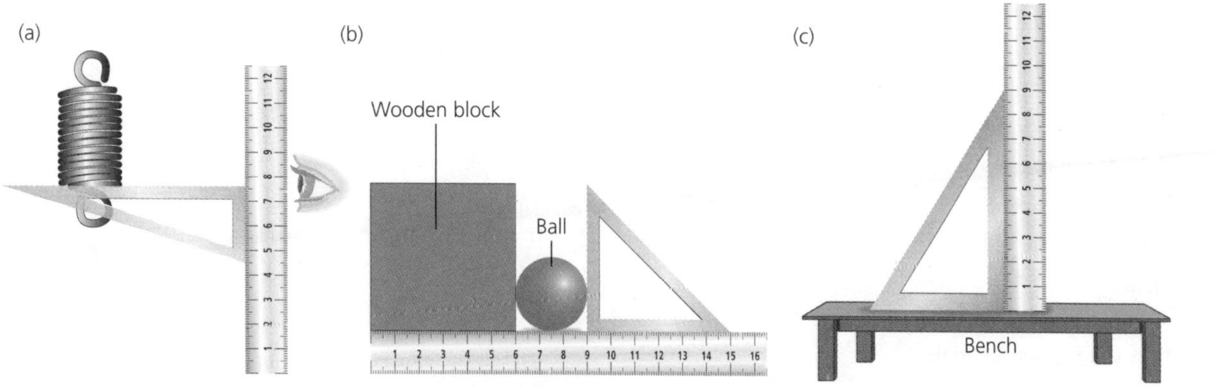

(a)   (b)   (c)

Wooden block

Ball

Bench

Figure 22

Make sure you keep your eye level, as shown in Figure 22a. Alternatively, to make sure that a clamped ruler is vertical in relation to the bench (Figure 22c), a plumb-line can be used — a small metal bob on a piece of string will hang vertically if suspended freely. This can be clamped next to the ruler.

Worked example

Determining the spring constant of a spring

You are asked to investigate the limit of proportionality and determine the spring constant of a spring. Describe how you would do this. Your answer should include:

- a list of apparatus
- a diagram of the experimental set-up
- an outline of the procedure (including safety considerations)
- what you could do to reduce the uncertainty in your measurements of length and extension

Answer

Apparatus: retort stand and clamp, metre rule, set-square, slotted masses and hanger, spring

Set-up: Figure 23 shows how the apparatus is set up.

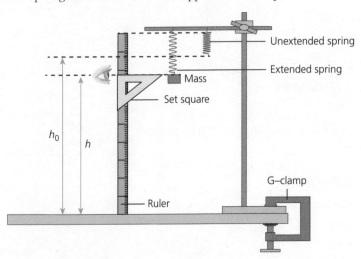

Figure 23

Safety considerations:

- Attach the masses securely to the spring and ensure that, should the masses fall, they will not fall on your feet. You may need to protect the floor with a crash pad.
- The clamp stand must be securely attached to the bench with a G-clamp.

Method:

- Secure the ruler as close to the spring as possible and use the set-square to ensure that the rule is vertical.
- Read the position of the bottom coil of the spring, using the set-square to ensure that the line between the spring and the ruler is horizontal.

→

Knowledge check 43

Explain how you could use a set-square to check that a beam is horizontal with respect to the bench.

Exam tip

Note the use of a set-square as an aid to taking the reading.

■ Ensure the spring is directly over the scale and align the eye directly above the scale to avoid parallax error.

■ Add the mass hanger and masses one at a time and read the new position of the bottom ring each time until the extension is no longer linear.

■ For each reading, calculate the extension of the spring by subtracting the original reading h_0 from the new reading h.

Using trigonometry to measure small angles

At best, a protractor can measure to 0.5° by interpolating between divisions. For an angle of, say, 5°, this would give rise to an uncertainty of 10%. By using trigonometric methods to measure small angles, we can significantly reduce this uncertainty. If you can measure the adjacent and opposite sides using a ruler, as shown in Figure 24, then the angle can be calculated using $\theta = \tan^{-1}(\text{opposite/adjacent}) = \tan^{-1}(h/l)$.

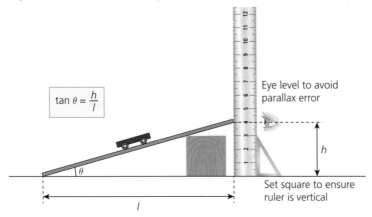

$\tan \theta = \dfrac{h}{l}$

Eye level to avoid parallax error

Set square to ensure ruler is vertical

Figure 24

Techniques for timing oscillations

Often, in experiments, the uncertainty in the reading is due not so much to the measuring instrument itself but rather to how it is used. There are a number of experiments at A-level that require you to time an object that is oscillating.

Precautions when timing oscillations:

■ Take care not to miscount oscillations — say 'nought' when starting to time.

■ Use a *fiducial marker* (i.e. a pointer) behind the object to help you judge when the oscillating object passes the centre/equilibrium position. The speed of the oscillator is fastest as it passes through the equilibrium position, so you should time from this point. If you try to time from the maximum displacement, the object will be moving very slowly and it will be very difficult to judge when it has actually reached the point of maximum displacement and is about to reverse direction.

Front view   **Side view**

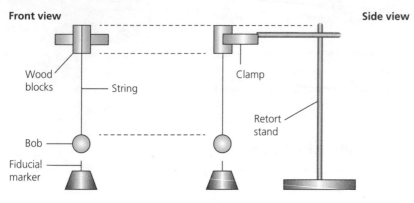

Figure 25

- Only start timing once the object has completed one or two oscillations. This will help you to start timing at the correct point. It will also ensure that you have not affected the motion of the object by giving it a small push at the start.
- Try to time at least 20 oscillations, if you can, in order to reduce the uncertainty. Obviously, if the system is heavily damped, then this will be difficult as the amplitude of the oscillations will decrease rapidly. In this case, take proportionately more repeat readings.

Worked example

Oscillations of a metre rule

A student makes some measurements of the time period T of the oscillations of a wooden metre rule in order to find a value for the Young modulus of the wood, using the arrangement shown in Figure 26. She finds in a textbook that the Young modulus E for the wood is given by the formula:

$$E = \frac{16\pi^2 ML^3}{wt^3 T^2}$$

where M is the mass suspended from the rule at a horizontal distance L from where the rule is clamped, and w and t are the width and thickness of the rule, respectively.

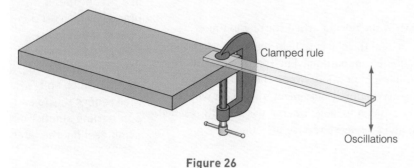

Figure 26

Knowledge check 45

A pendulum bob is released a vertical distance of 10 cm above its equilibrium position. Calculate the maximum speed of the bob as it swings through the lowest point of its motion.

She plans to use:

- **two 100 g slotted masses** attached with elastic bands to either side of the rule, near the end, to make L as long as possible — with the slots in the masses perpendicular to the rule so that she can read the scale to determine the position of the centre of mass
- **a digital stopwatch** to time the oscillations — one that can record to a resolution of 0.01 s, which she considers to be more than adequate for timings that are likely to be of the order of 10 s
- **vernier callipers** to measure the width and thickness of the rule, at different points along its length — the callipers can read to a resolution of 0.1 mm
- a pin secured to the metre rule with a small piece of modelling clay to act as a **fiducial mark** to help her judge the centre of the oscillations

She records the following results:

$$M = 200\,\text{g}$$

$$L = 937 - 40 = 897\,\text{mm}$$

$$w = 28.5,\ 28.0,\ 28.6,\ 28.9\,\text{mm; average } 28.5\,\text{mm}$$

$$t = 6.6,\ 6.7,\ 6.6,\ 6.6\,\text{mm; average } 6.6\,\text{mm}$$

$$20T = 8.59,\ 8.54,\ 8.18,\ 8.61\,\text{s; average } 8.48\,\text{s}$$

a Explain why a valu\e of 8.58 s for the average time should be used rather than the value stated by the student.

b Show that these data give a value of the order of $10^{10}\,\text{Pa}$ for the Young modulus of the wood.

c Explain the number of significant figures to which the final value should be given.

d Suggest a suitable uncertainty for the value of the period T.

e Discuss whether the selection of vernier callipers, which could only be read to a precision of 0.1 mm, to measure the width and thickness of the rule, was an *appropriate strategy*. Justify your answer by estimating the percentage uncertainty that these measurements introduce into the value obtained for the Young modulus.

f To what extent would using a micrometer screw gauge, or digital callipers, reading to 0.01 mm, reduce the percentage uncertainty in the value obtained for the Young modulus?

Answer

a Looking at the four values for $20T$, it would appear that the 8.18 reading is anomalous, as it is about 0.5 s smaller than the other values (suggesting only 19 oscillations may have been counted). This value should therefore be ignored (with a note to this effect) when determining the average value.

b Care needs to be exercised with units — all measurements need to be converted to SI units:

$$M = 0.200\,\text{kg}$$

$\rightarrow$

Exam tip

You are expected *to comment on experimental design*, so, when writing a plan, always state the *reason* for your choice of each measuring instrument or for a particular technique that you use.

Exam tip

Always record measurements with appropriate units and record *all* your measurements, not just the average values.

Exam tip

Results that are clearly anomalous should be ignored when calculating the average of several values, and the fact that you have done this should be stated. Note that one advantage of taking repeat readings is that an anomalous reading can be spotted and allowed for.

$$L = 0.897\,\text{m}$$

$$w = 28.5 \times 10^{-3}\,\text{m}$$

$$t = 6.6 \times 10^{-3}\,\text{m}$$

$$T = 8.58\,\text{s} \div 20 = 0.429\,\text{s}$$

so

$$E = \frac{16\pi^2 ML^3}{wt^3 T^2} = \frac{16\pi^2 \times 0.200\text{kg} \times (0.897\text{m})^3}{28.5 \times 10^{-3}\,\text{m} \times (6.6 \times 10^{-3}\,\text{m})^3 \times (0.429\text{s})^2}$$

$$= 1.5 \times 10^{10}\,\text{Pa} \sim 10^{10}\,\text{Pa}$$

c The value should be quoted to two significant figures, 1.5×10^{10} Pa, because this is the number of significant figures of the least precise measurement — the thickness of the rule, 6.6 mm, has been measured to only 2 sf.

d Although the time taken for 20 oscillations has been recorded to a precision of 0.01 s, the range of values is $(8.61 - 8.54)\,\text{s} = 0.07\,\text{s}$. This is probably slightly less than the human reaction time. A more realistic uncertainty in $20T$ would be 0.1 s. This would give an uncertainty for the value of T of $0.1 \div 20 = 5$ ms. We can then say $T = 0.429 \pm 0.005\,\text{s}$.

e The measurements recorded for the width were:

$w = 28.5, 28.0, 28.6, 28.9$ mm; average 28.5 mm

The range of values is therefore $(28.0 - 28.9)\,\text{mm} = 0.9\,\text{mm}$.

If we use half the range (0.45 mm) as our uncertainty, we get:

$$\%\,\text{uc in}\,w = \frac{0.45\,\text{mm}}{28.5\,\text{mm}} \times 100\% = 1.6\%$$

Using a vernier scale reading to a precision of 0.1 mm to measure a thickness of 6.6 mm means that there is a percentage uncertainty in t of

$$\frac{0.1\,\text{mm}}{6.6\,\text{mm}} \times 100\% = 1.5\%$$

As the term t^3 occurs in the formula for the Young modulus, the percentage uncertainty introduced by this measurement will be $3 \times 1.5\% = 4.5\%$.

The combined contribution of the measurements for width and thickness is therefore:

$$\%\,\text{uc} = 1.6\% + 4.5\% = 6.1\%$$

This is a not insignificant uncertainty. A vernier only reading to 0.1 mm to measure the thickness was not a good choice.

f If an instrument reading to 0.01 mm had been used, the percentage uncertainty in the thickness would have been reduced by a factor of 10, giving an uncertainty due to this measurement of less that 0.5%, which is perfectly acceptable. However, due to the relatively large spread of values in the measurement of the width, using a more sensitive instrument would not have had much effect on the previously calculated uncertainty of 1.6%. The overall uncertainty would therefore be about 2%, which is nevertheless a considerable improvement.

Knowledge check 46

A student determined the refractive index of glass five times: 1.44, 1.42, 1.46, 1.56, 1.43. What should the student quote as the mean value for refractive index, and why?

Using data loggers

Data loggers record data automatically and can record data with very small intervals or over long periods of time. They can record thousands of measurements per second, with the data output displayed on a computer as a graph or in a table. The number of readings taken per second is called the *sample rate*. Sensors are available to enable a wide range of measurements to be made, e.g. force, current, potential difference, magnetic flux density and temperature. The data logger is able to measure several variables simultaneously and so can, for example, record how force varies with time. This graph could then be used to calculate the impulse by determining the area underneath the line.

Worked example

In an experiment to investigate how the current changes with time as a capacitor is charged, a data logger attached to a current sensor is used instead of a digital ammeter and stopwatch. State three advantages of using a data logger in this way.

Answer

You could state:

- eliminates reaction time errors
- automatically takes current and time readings simultaneously
- updates the display much quicker than a standard ammeter
- high sample rate and so records many readings of current per second
- immediately measures the initial current the instant the experiment is started
- the data are stored in a computer and so can be easily processed and analysed (e.g. the gradient or area under a graph calculated)

Worked example

A magnet is dropped through a coil of many turns. This is connected to a voltage sensor, which records the induced emf as the magnet falls through the coil (Figure 27a). Typical results are shown in Figure 27(b).

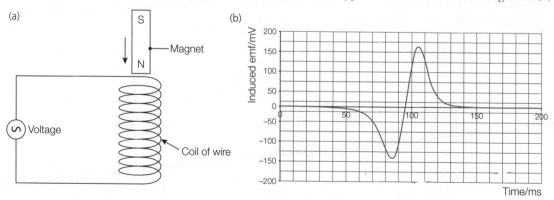

Figure 27

Discuss the advantages of using a data logger to collect these data.

→

Answer

You should include the following points:

- The event happens over about 100 ms, which does not give enough time for the data to be recorded manually.
- The high sample rate of the data logger enables the induced emf to be measured hundreds of times during this short time interval.
- This produces enough data for a graph to be plotted showing how the induced emf varies with time.
- If the data logger is connected to a computer, the graph can be shown directly.

Analysing results

You will be expected to:

- *plot and interpret graphs*
- *process and analyse data using appropriate mathematical skills as exemplified in the mathematical appendix of the specification*

Working through the *Maths and units* section of this student guide will give you plenty of practice in developing your mathematical skills. You should note that more advanced mathematical skills are required for A-level, which mainly means ensuring that you are confident in handling exponential functions and drawing and analysing logarithmic graphs.

Plotting a graph is a key first step in analysing your results. All graphs should be drawn with sensible scales and with appropriately labelled axes, including correct units. Points should be plotted accurately and a straight line, or smooth curve, of best fit drawn.

What graph should I draw?

Scientists carry out experiments in order to find an equation that relates the independent (x) and dependent (y) variables. If the graph obtained is a straight line of gradient m and a y-intercept of c, the relationship can be written in the form of $y = mx + c$.

If the graph is a straight line *passing through the origin* ($c = 0$), then we can say that variable y is directly proportional to variable x ($y = mx$). For example, in an experiment to measure the resistance of a wire, the graph shown in Figure 28 was obtained.

Exam tip

It is important that you learn how to rearrange a formula so that you can get a straight-line graph and be able to identify the variables and the terms for the gradient and intercept (if there is one).

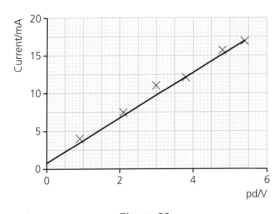

Figure 28

The graph does not quite pass through the origin, which is probably due to a systematic error in the measurements. This could be due to a zero error in the ammeter or voltmeter, or possibly some contact resistance at the terminals where the wire is connected into the circuit.

If the line is a straight line but clearly does not pass through the origin, then we can say that there is a *linear relationship* between the two variables, but this is *not* a proportional relationship.

Many experiments result in graphs that are curves. Perhaps there is an inversely proportional relationship between the variables or a power law? Maybe the relationship is exponential?

Worked example

Oscillations of a metre rule

The student's teacher in the example on page 38 suggests that a better value for the Young modulus might be obtained if measurements were taken with different masses and a suitable graph was plotted. The relevant equation is:

$$E = \frac{16\pi^2 ML^3}{wt^3T^2}$$

a For a given metre rule, this equation for the Young modulus contains three variables. State what they are and, following the teacher's suggestion, state which one must be kept constant? How would you ensure this?

b Explain what should be plotted in order to get a linear graph.

c Show how the graph could be used, together with the rest of the experimental data, to get a value for the Young modulus.

d Suggest why a graph is likely to reduce the uncertainty in the value obtained for the Young modulus.

Answer

a For a given rule, the three variables are:
- the suspended mass M
- the period T
- the length of suspension L

Of these, L should be kept constant and T should be found for different values of M.

To ensure L remains constant:
- the masses should be secured tightly
- the distance should be checked before each timing

b From the equation:

$$E = \frac{16\pi^2 ML^3}{wt^3T^2} \implies T^2 = \frac{16\pi^2 l^3}{Ewt^3} \times M$$

Therefore, a graph of T^2 against M should be plotted.

→

Knowledge check 49

The braking distance of a car is directly proportional to its mass. If a 2000 kg car has a braking distance of 120 m for a particular speed, calculate the braking distance required for a 1400 kg car travelling at the same speed.

Exam tip

You will lose marks if the line is forced through the origin or if a straight line is drawn when the data suggest a curve.

Exam tip

Practice so that you are confident in arranging complex formulae to give straight-line graphs of the form $y = mx + c$ and in determining which of the terms are variables and which are constant.

c The graph should be a straight line through the origin:

$$\text{gradient} = \frac{16\pi^2 L^3}{Ewt^3} \quad \Rightarrow \quad E = \frac{16\pi^2 L^3}{\text{gradient} \times wt^3}$$

d Plotting a graph would:
   ■ give an average of a number of readings, and so
   ■ reduce both random and systematic errors

Lines of best fit

Before you draw a line of best fit, think about the underlying physics to help you decide whether the line should be straight or curved. Perhaps there is a law that applies to these data (e.g. $V = IR$)?

Use these guidelines to ensure that you draw the best possible line of best fit. Examiners expect lines to be drawn carefully (i.e. thin and straight or smooth).

■ Try to make sure there are as many points on one side of the line as the other.
■ The line should pass very close to the majority of plotted points.
■ Are there any obviously anomalous results?
■ Are there uncertainties in the measurements? The line of best fit should fall within error bars if those are drawn.
■ Use a thin, sharp pencil to draw the line of best fit.

Knowledge check 50

How would you check graphically whether the experimental results fit the following equations?

a $mgh = \tfrac{1}{2}mv^2$ for variables h and v

b $T = 2\pi\sqrt{\dfrac{l}{g}}$ for variables T and l

Worked example

How *not* to plot graphs!

Look at the graphs in Figure 29 and decide what is wrong with each one.

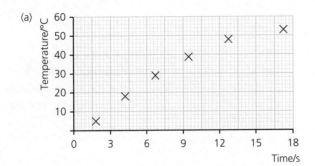

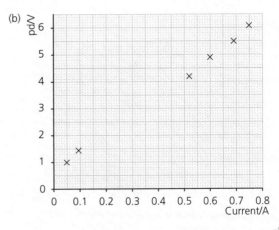

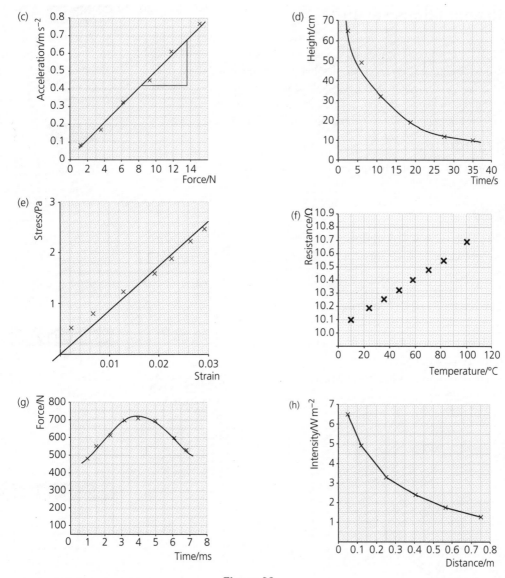

Figure 29

Answer

a Awkward scale divisions — in particular, avoid multiples of 3 as interpolation is difficult.

b No readings between 0.1 A and 0.5 A — a poor distribution of readings makes it difficult to plot a line of best fit.

c The triangle used to calculate the gradient is too small — this would produce large percentage uncertainties in the values of Δx and Δy and therefore in the gradient.

d Poorly drawn curve of best fit — a smooth curve is required (this line is 'spidery').

e The line has been forced through the origin, and so is not the 'best fit'.

f A sharp pencil was not used to plot points, making them very thick — precision is lost.

g Poor y-scale — the points do not fill more than half the y-axis, so the scale could have been doubled.

h A line has been drawn simply joining the points — a *smooth* curve of best fit should have been drawn.

Content Guidance

Dealing with anomalous results

You may find that some results do not fit the general pattern and lie a long way from the line of best fit. Circle these and do not include them in your line of best fit. You should make it clear that you have done this. Try to think what may have caused them. Did a different student collect that piece of data? Was a different measuring device used? Did the external conditions change? If there is time, this point should be repeated. You should also exclude any anomalous values when calculating the mean of repeats.

Calculating gradients

A gradient often represents an important quantity, as discussed on page 14 in the section on mathematical requirements.

Worked example

Determining the resistance of a component

The circuit shown in Figure 30 is set up to determine the resistance of a component.

The pd (V) across the component and the current (I) in it were measured for a range of voltages and the results shown in Table 7 were obtained.

a Plot a graph of V on the y-axis against I on the x-axis.

b Comment on the graph you obtain.

c Use your graph to determine a value for the resistance of the component.

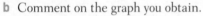

Figure 30

Answer

a

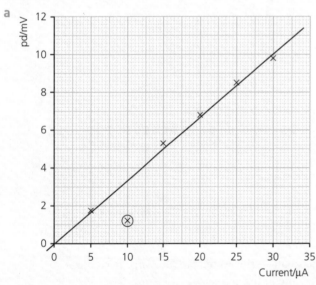

Figure 31

b The graph is a straight line through the origin. The point (10.0 μA, 1.27 mV) is an anomalous result, and so is excluded from the line of best fit.

c $R = V/I = \text{gradient} = \dfrac{(10.0 - 0.0) \times 10^{-3}}{(30.0 - 0.0) \times 10^{-6}}$

$R = 333\,\Omega = 330\,\Omega$ (2 sf)

Table 7

I/μA	V/mV
5.0	1.75
10.0	1.27
15.0	5.10
20.0	6.72
25.0	8.35
30.0	9.88

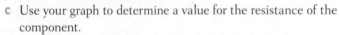

> **Exam tip**
>
> If you decide to exclude any anomalous result, you should state that you have done so.

> **Exam tip**
>
> You must use a large triangle with a base as large as possible (normally at least 8 cm) and remember to include the units in your final calculation of a gradient.

> **Knowledge check 51**
>
> What is the resistance of a resistor if the pd across it is 4.97 mV when the current in it is 22.6 μA?

> **Knowledge check 52**
>
> a State Ohm's law.
> b Sketch a graph of the data you would use to check whether a component obeyed Ohm's law.
> c How could you tell whether the component was ohmic?
> d How could you use your graph to determine the resistance of the component?

Worked example

Calculating the spring constant

Explain how you would use the data for the spring constant experiment on page 36 to determine the limit of proportionality and the spring constant.

Answer

- For each extension, calculate the force applied using $W = mg$.
- Plot a graph of force against extension. Note that, contrary to normal practice, we are plotting the dependent variable (extension) on the x-axis. This is so that the gradient gives us the spring constant — see the last bullet point below.
- Read off the point at which the graph begins to curve — the limit of proportionality.
- For the linear region, determine the gradient by drawing a large triangle — this gives the spring constant.

Error bars and uncertainties in the gradient

The uncertainty in a measurement can be shown on a graph as an error bar. If there is an uncertainty in both the quantities y and x, then instead of an error bar you would have an error rectangle.

If you are plotting a log graph, then you have to calculate the logarithm of the maximum and minimum possible values for each data point and determine the error bars using these values.

The uncertainty in each measurement can be determined from the resolution of the data or it is sometimes quoted as a percentage. Use the following guidelines to help you draw correct error bars.

If there are repeat readings:
- plot the data points at the mean values
- calculate the range of the data, ignoring any anomalies
- add error bars with lengths equal to half the range on either side of the data points
- if you know the absolute uncertainty, plot the error bars based on this straight away

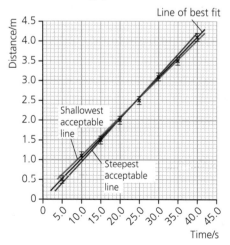

Figure 32

Exam tip

When plotting graphs, remember to plot the correct quantities on each axis. *Force against time* means that *force* would go on the y-axis and *time* on the x-axis.

Knowledge check 53

Determine the uncertainty in $\ln T$ that should be used to plot error bars if $T = (8.2 \pm 0.2)\,\text{N}$.

To determine the uncertainty in a gradient, two gradients should be drawn on the graph, as shown in Figure 32:
- the line of 'best fit' and
- *either* the steepest (shown in blue) *or* the shallowest (shown in red) gradient lines that can be drawn through the data points (or through the error bars if these are shown)
- the gradient of each line should be found, then

percentage uncertainty in gradient =

$$\frac{\text{(highest (or lowest) possible gradient} - \text{gradient of best fit)}}{\text{gradient of line of best fit}} \times 100$$

Worked example

Determining the capacitance of a capacitor

A student is checking the value of a capacitor by charging it to known potentials and measuring the corresponding charge stored in the capacitor using the circuit shown in Figure 33. The coulomb meter has a tolerance of ±10%.

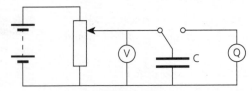

Figure 33

a How could the student maximise the precision of the meter readings?

b The student recorded the following data:

Potential difference/V	Charge stored/µC
0.00	0
1.20	377
2.40	757
3.60	1167
4.80	1505
6.00	1904
7.20	2275

Plot a suitable graph of the data, using error bars to indicate the uncertainty in the meter readings, and use your graph to determine the capacitance of the capacitor.

c Estimate the percentage uncertainty in the value of your gradient and hence the uncertainty in your value for the capacitance. Discuss the extent to which this confirms the stated value for the capacitor as 330 µF ± 10% tolerance.

Answer

a For the digital meters, the range that gives the largest number of decimal places should be chosen. The range of the meter should not be changed during the experiment as this could alter the uncertainty of the meter and may give slightly different readings.

→

b The graph is shown in Figure 34.

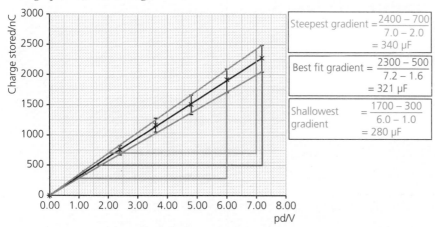

Steepest gradient $= \dfrac{2400 - 700}{7.0 - 2.0}$
$= 340\,\mu F$

Best fit gradient $= \dfrac{2300 - 500}{7.2 - 1.6}$
$= 321\,\mu F$

Shallowest gradient $= \dfrac{1700 - 300}{6.0 - 1.0}$
$= 280\,\mu F$

Figure 34

Best fit gradient = capacitance = $321\,\mu F$ (see graph).

c %uc in gradient $= \dfrac{\frac{1}{2}(\text{steepest} - \text{shallowest})}{\text{best fit}} \times 100\%$

$= \dfrac{\frac{1}{2}(340 - 280\,\mu F)}{321\,\mu F} \times 100\% = 9\%$ (see graph)

Absolute uc in capacitance = 9% of $321\,\mu F = 29\,\mu F$

Capacitance = $321 \pm 29\,\mu F$ or, more realistically, $320 \pm 30\,\mu F$ to 2 sf.

Within the tolerance of the coulomb meter, the value of $320\,\mu F$ for the capacitance is in agreement with the stated value.

Exponential changes and log graphs

(A-level only) At A-level you will need to be familiar with exponential changes. Common examples include radioactive decay, the decrease in amplitude of a damped oscillator and the discharge of a capacitor through a resistor.

A good example of an exponential relationship is shown in an experiment to investigate how the length of a jelly fibre affects the amount of light that is absorbed. A phototransistor is used to detect the light. This produces a voltage that is directly proportional to the amount of light falling on it.

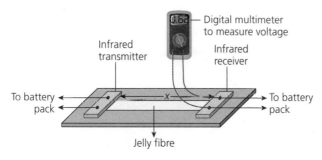

Figure 35

Exam tip

Your graph should have

- a sensible scale
- axes labelled with units
- 10% error bars plotted
- best fit line
- gradient from large triangle using best fit line
- steepest and shallowest gradients

Knowledge check 54

(A-level only) Sketch a graph of $\ln A$ against t for radioactive decay and explain how you could find the half-life from the gradient.

Exam tip

Remember that when dealing with exponential functions you *must* take *natural* logarithms (i.e. to base 'e') using the 'ln' function on your calculator.

The following data were obtained in such an experiment:

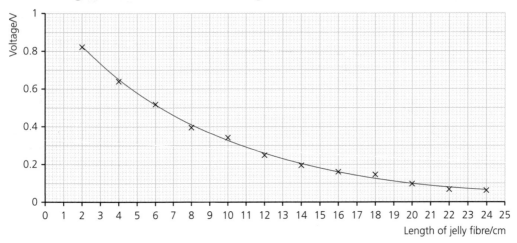

Figure 36

The exponential shape of the curve suggests that the intensity of light passing through the fibre might decrease exponentially with length according to the relationship:

$$I = I_0 e^{-\mu x}$$

where I = intensity of light, μ = absorption coefficient of jelly and x = length of jelly.

Taking natural logarithms ('ln') on both sides of the equation gives:

$$\ln I = \ln I_0 - \mu x$$

Since the phototransistor voltage V is proportional to the intensity I:

$$\ln V = \ln V_0 - \mu x$$

Comparing with $y = mx + c$, we can see that, if $\ln V$ is plotted on the y-axis against x on the x-axis, we should expect a straight-line graph of gradient $= -\mu$ and y-intercept $= \ln V_0$ as shown in Figure 37. This is called a log-linear graph.

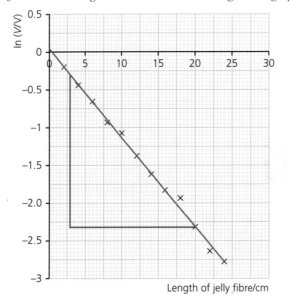

Length of jelly fibre/cm

Figure 37

Exam tip

You can also test for exponential changes mathematically using the 'constant ratio test'. Applying this to the data shown in Figure 36:

  $0.64/0.82 \approx 0.78$

  $0.52/0.64 \approx 0.81$

  $0.40/0.52 \approx 0.77$

These ratios are all within a few per cent of each other, so within experimental limits we can say that the data show an exponential decrease in voltage with an increase in length.

Worked example

a Use the graph in Figure 37 to determine the absorption coefficient, μ.
b Use this graph to find the value of V_0.
c Determine the length of fibre needed to reduce the intensity to 15% of its initial amount.

Answer

a Taking the points (3.00, −0.30) and (20.0, −2.30) gives:

$$\text{gradient} = \frac{-2.30 - (-0.30)}{(20.0 - 3.0)\text{cm}} = -0.12\,\text{cm}^{-1}$$

The absorption coefficient is therefore $0.12\,\text{cm}^{-1}$.

b y-intercept $= 0.04 \Rightarrow V_0 = e^{0.04} = 1.0\,\text{V}$
c Substituting $I = 0.15 I_0$ and $\mu = 0.12\,\text{cm}^{-1}$ into the equation $I = I_0 e^{-\mu x}$

$$0.15\,I_0 = I_0 e^{-0.12x} \Rightarrow \ln(0.15) = -0.12(x/\text{cm}) = -1.897$$

$$(x/\text{cm}) = -1.897/(-0.12)$$

$$x = 16\,\text{cm}$$

Log graphs can also be used to investigate relationships of the form:

$$y = Ax^n$$

where A and n are constants.

It would take a long time to find n using trial and error, but if we take logs of both sides (either to base '10' or to base 'e') the equation becomes

$$\log y = \log A + n \log x \text{ or } \ln y = \ln A + n \ln x$$

A graph of $\ln y$ against $\ln x$ will give a straight line. Comparing this with the equation of a straight line, $y = mx + c$, gives

$$\ln y = \ln A + n \ln x$$

$$y = \quad c \ + \quad mx$$

i.e. gradient $m = n$ and intercept $c = \ln A$.

Evaluating results and drawing conclusions

You will be expected to *evaluate results and draw conclusions with reference to measurement uncertainties and errors*.

Exam questions may ask you to discuss to what extent your results confirm a particular relationship between two variables. You may also be asked to comment on accuracy and precision, and to discuss how the errors and uncertainties in the experiment could be reduced. The following worked examples give you practice at applying your knowledge of *planning, making measurements, recording data, analysing results* together with *evaluating results and drawing conclusions*.

Exam tip

Always include the units when determining the gradient of any graph.

Exam tip

Always try to check the validity of your answer if possible. Here you should look back at the data and see whether the voltage has dropped to 15% of its original value for a length of 16 cm.

Exam tip

For an exponential function, you *must* use natural logs ('ln'). For investigating a power, you can use either 'ln' or 'log'. If you always use 'ln', you cannot go wrong.

Exam tip

Remember, the intercept is $\ln A$ (or $\log A$) and so $A = e^{y\text{-intercept}}$ (or $10^{y\text{-intercept}}$).

Reminder: the percentage difference between two experimental values is given by:

$$\% \text{ difference} = \frac{\text{difference between the values}}{\text{average of the two values}} \times 100\,\%$$

The percentage difference between an experimental value and a stated or known value is given by:

$$\% \text{ difference} = \frac{\text{difference between the values}}{\text{stated value}} \times 100\,\%$$

Worked example

Investigating resonance

A student plans to investigate how the frequency of vibration of air in a conical flask depends on the volume of air in the flask. She blows directly into the neck of the flask and listens to the sound of the air vibrating. She then pours water into the flask until it is approximately half filled and blows into the flask as before. She notices that the pitch (frequency) of the vibrating air is higher when the flask is half full of water.

She thinks that there might be a relationship between the natural frequency of vibration f of the air in the flask and the volume V of air in the flask of the form $f \propto V^n$, where n is a numerical constant.

The student sets up the arrangement shown in Figure 38.

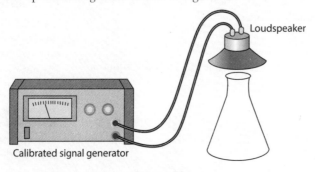

Figure 38

With the flask empty, she increases the frequency of the signal generator until the air in the flask vibrates very loudly. She repeats this for increasing volumes of water in the flask.

a Explain why this happens at the natural frequency of vibration of the air.

b Describe a technique by which the uncertainty in determining this frequency could be reduced.

c Describe how she could vary, and measure, the volume of air in the flask.

d Explain how plotting a graph of $\ln f$ against $\ln V$ would enable her to test whether $f \propto V^n$ and would enable her to find a value for n.

e The student obtained the data shown in the table. →

a A student determines the spring constant for a spring by two different methods — by measuring the extension for different loads and by timing vertical oscillations. Values of $26.7\,\text{N}\,\text{m}^{-1}$ and $28.3\,\text{N}\,\text{m}^{-1}$ are obtained, respectively. What is the percentage difference between these two values?

b The student also combines the results of the two experiments to find a value for g of $9.71\,\text{m}\,\text{s}^{-2}$. By what percentage does this value differ from the accepted value for g?

V/cm³	f₁/Hz	f₂/Hz	f/Hz
554	217	221	219
454	243	241	242
354	271	277	274
254	325	323	324
204	363	359	361
154	412	418	415

Calculate values of $\ln(V/cm)$ and $\ln(f/Hz)$ from the data in the table and add those values to the table. Then plot a graph of $\ln(f/Hz)$ against $\ln(V/cm)$ and use your graph to determine a value for n.

f Explain **qualitatively** whether your value for n is consistent with the student's initial observations.

Answer

a When the frequency of the signal generator, i.e. the frequency of vibration of the loudspeaker, is equal to the natural frequency of vibration of the air inside the flask, resonance occurs. Energy is transferred with maximum efficiency from the loudspeaker to the air in the flask, and so this air vibrates very loudly.

b The frequency should be increased gradually until resonance occurs. This frequency should be recorded. The frequency should then be increased beyond the resonant frequency and gradually reduced until once again resonance is detected. This frequency should also be recorded and the average of the two values should be taken as the resonant frequency.

c A measuring cylinder is required, then:
- this is used to fill the flask to the top with water, so that the volume V_f of the flask can be determined
- the flask is then emptied
- a known volume v of water is now added to the flask
- the volume of air will therefore be $V = V_f - v$
- further volumes of water are added to give different volumes V of air in the flask

d The proposed equation is $f \propto V^n$, or $f = kV^n$. If we take logarithms ('ln') on both sides of the equation, we get

$$\ln(f) = n\ln(V) + \ln(k)$$

So, if a graph of $\ln(f)$ against $\ln(V)$ is plotted, this should be a straight line with gradient equal to the constant n and intercept on the y-axis equal to $\ln(k)$.

e The values are shown in the table.

V/cm³	f/Hz	ln(V/cm³)	ln(f/Hz)
554	219	6.32	5.39
454	242	6.12	5.49
354	274	5.87	5.61
254	324	5.54	5.78
204	361	5.32	5.89
154	415	5.04	6.03

→

> **Exam tip**
>
> Note that the technique should be described in detail — an answer such as 'The reading was repeated and the average taken' is not sufficient. This technique can also be applied to focusing an image with a lens.

Your graph should be a straight line of negative slope, as in Figure 39.

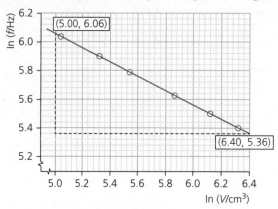

Figure 39

Note that:

- the points should occupy at least half the graph paper in each direction, which means that the scales should **not** start at the origin
- the axes should be labelled exactly as shown: $\ln(f/\text{Hz})$ on the y-axis and $\ln(V/\text{cm})$ on the x-axis
- the points should be plotted to a precision of at least half a square
- a thin, straight line of best fit should then be drawn

The constant n is the gradient of the graph. Taking a large triangle:

$$\text{gradient} = n = \frac{6.06 - 5.36}{5.00 - 6.40} = \frac{+0.70}{-1.40} = -0.50$$

(do not forget the minus sign)

f A negative value of n means that, as the volume of air in the flask gets less, the frequency of vibration (pitch) of the air gets higher, which is consistent with the student's initial observation.

A value of -0.50 suggests that $n = -\frac{1}{2}$, meaning that:

$$f \propto V^{-1/2} \quad \text{or} \quad f \propto \frac{1}{\sqrt{V}}$$

In other words, the frequency of vibration is inversely proportional to the square root of the volume of air.

Knowledge check 56

The period T of vertical oscillations of a mass m on a spring of spring constant k is given by

$$T = 2\pi\sqrt{\frac{m}{k}}$$

a Write down the expression obtained by taking natural logarithms on both sides of this equation.

b State the value of the gradient when $\ln(T/\text{s})$ is plotted against $\ln(m/\text{kg})$.

c The y-intercept is found to be 0.19. What value does this give for the spring constant k?

■ Required practicals

Required practical 1

Investigating the variation of the frequency of stationary waves on a string with length, tension and mass per unit length of the string

Essential theory

If the wire in Figure 40 is plucked, it will vibrate at its fundamental frequency, with nodes at each end and one antinode at its centre. The distance l between the nodes is $\frac{1}{2}\lambda$ and so $\lambda = 2l$.

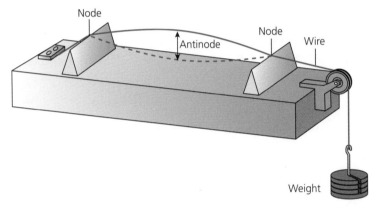

Figure 40

For a wire or string of mass per unit length μ under a tension T, the fundamental frequency f is given by the equation:

$$f = \frac{1}{2l}\sqrt{\frac{T}{\mu}}$$

We can see that f depends on *three* variables: l, T and μ. We must therefore keep two of these variables constant whilst the relationship between f and the third variable is investigated.

Planning

For a given wire, μ will be constant. Therefore, we can either keep l constant and investigate how f depends on T, or keep T constant to see how f depends on l. Two ways of doing this are shown in Figure 41.

Knowledge check 57

Show that the fundamental frequency of a wire having a length of 50 cm and a mass per unit length of 2 g m^{-1} under a tension of 10 N is about 70 Hz.

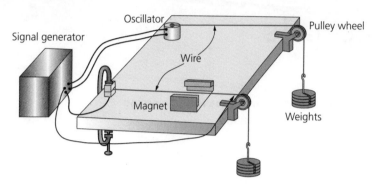

Figure 41

- If the *length is kept constant*, the frequency can be adjusted so that the wire vibrates at its fundamental frequency for a series of weights added to the wire.
- Likewise, if the wire is under the *tension of a fixed load*, the fundamental frequency can be found for different lengths of the wire.
- Appropriate graphs can then be plotted to test the relationship.

Safety

Some basic precautions should be observed — see part **a** of the worked example below.

Worked example

Table 8 shows a typical set of measurements for a steel wire of length $l = 0.500\,\text{m}$.

Table 8

Mass/kg	T/N	Frequency/Hz	Average f/Hz	$f^2/10^3\,\text{Hz}^2$
0.50	4.9	46,50	48	
1.00	9.8	69,63	66	
1.50	14.7	85,77	81	
2.00	19.6	90,98	94	
3.00	29.4	111,117	114	
4.00	39.2	137,129	133	

a State any safety precautions you would adopt when doing this experiment.

b Explain how a graph of f^2 on the y-axis against T on the x-axis can be used to show the relationship between T and f.

c Add values for f^2 to Table 8 and plot a graph of f^2 against T. What does your graph enable you to deduce about the relationship between T and f?

d Use the gradient of your graph to show that the mass per unit length μ of the wire is about $2\,\text{g}\,\text{m}^{-1}$.

e Describe how you would check your value for μ by direct measurement.

f Sketch the graph you would plot to investigate the relationship between f and l for a fixed load.

g Explain briefly how you would investigate the relationship between f and μ. You should include a sketch of the graph you would plot.

→

Exam tip

It is good practice to determine the frequency by increasing it until a large vibration is observed and recording this value. Then increase the frequency a bit more and gradually reduce it again until a large vibration is once more observed, giving a second reading. The two readings can then be averaged.

Answer

a The following precautions should be observed:

- Place a suitable 'crash pad' under the weights to prevent them hitting the floor directly if the wire breaks.
- Do not place your feet under the weights (for the same reason!)
- Wear goggles for protection in case the wire breaks.

b Squaring the formula $f = \dfrac{1}{2l}\sqrt{\dfrac{T}{\mu}}$ on both sides gives us $f^2 = \dfrac{T}{4l^2\mu}$. Therefore, a graph of f^2 against T should be a straight line through the origin with a gradient $\dfrac{1}{4l^2\mu}$.

c The values for $f^2/10^3\,\mathrm{Hz}^2$ are: 2.3, 4.4, 6.6, 8.8, 13.0 and 17.7. Your graph of f^2 against T should be a straight line *through the origin* and the point (40 N, $18 \times 10^3\,\mathrm{Hz}^2$). This shows that $f^2 \propto T$ or $f \propto \sqrt{T}$ as in the equation.

d From answer **b** the gradient is $\dfrac{1}{4l^2\mu}$.

From answer **c** its value is $\dfrac{(18.0 - 0.000) \times 10^3\,\mathrm{s}^{-2}}{(40.0 - 0.00)\mathrm{N}} = 450\,\mathrm{N}^{-1}\mathrm{s}^{-2}$.

$\mu = \dfrac{1}{\text{gradient} \times 4l^2} = \dfrac{1}{450\,\mathrm{N}^{-1}\mathrm{s}^{-2} \times 4 \times 0.250\,\mathrm{m}^2}$

$= 2.2 \times 10^{-3}\,\mathrm{kg\,m}^{-1} = 2.2\,\mathrm{g\,m}^{-1}$

e As μ is only $2\,\mathrm{g\,m}^{-1}$, the mass of, say, 5 m of wire should be found ($\approx 10\,\mathrm{g}$) using a balance of sensitivity 0.1 g or better.

f For constant T, we need to plot f against $1/l$ to test whether $f \propto 1/l$. This is shown in Figure 42.

g To investigate how f is related to μ, the fundamental frequency should be found for different wires (i.e. different μ), keeping l and T the same for each wire. A graph of f^2 against $1/\mu$ should then be plotted, as in Figure 43.

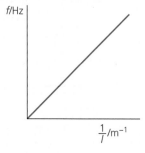

Figure 42

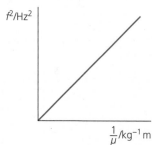

Figure 43

Exam tip

You must state that the graph *passes through the origin* when testing for a proportional relationship. No marks will be awarded for just saying 'a straight line'.

Knowledge check 58

Show that the units of $1/(\mathrm{N}^{-1}\mathrm{s}^{-2} \times \mathrm{m}^2)$ are equivalent to $\mathrm{kg\,m}^{-1}$.

Required practical 2

Investigation of interference effects to include the Young's slit experiment and interference by a diffraction grating

Essential theory: diffraction gratings

When monochromatic light, such as a laser beam, is shone through a diffraction grating, a pattern of lines is seen. There is a central bright maximum, with further maxima of decreasing intensity either side (Figure 44 a). This is represented diagrammatically in Figure 44 (b). These effects can be used to determine the wavelength of laser light.

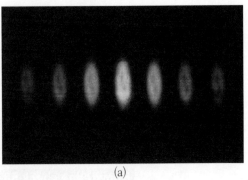

(a)

$n = 3$
$n = 2$
$n = 1$
$n = 0$
Monochromatic light
$n = 1$
$n = 2$
Grating
$n = 3$

(b)

Figure 44

The angle θ of a particular maximum from the central maximum is given by the equation

$$n\lambda = d \sin \theta$$

where n is the 'order' of the maximum (i.e. $n = 1$ is the first maximum on each side of the central maximum, and so on), d is the separation of the lines of the grating and λ is the wavelength of the light.

Planning

The experimental arrangement is shown in Figure 45. The light source is a laser emitting green light.

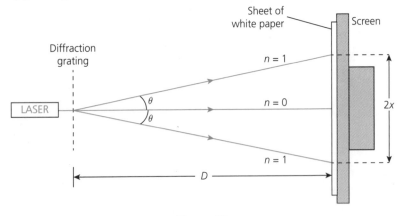

Figure 45

Knowledge check 59

Show that the angle of diffraction of the first order for light of wavelength 550 nm transmitted through a grating of spacing 300 lines per mm is about 10°.

You should proceed as follows:

- Adjust the distance D between the grating and the screen until a number of orders can be seen, sufficiently far apart for their separation to be measured with a millimetre scale.
- Measurements should *not* be attempted directly — the centres of the maxima should be marked on a piece of paper secured to the front of the screen and then processed after the laser has been switched off (see 'Safety' below).
- Measure the distance apart, $2x$, between the maxima of each order and then determine θ for each order using $\tan\theta = x/D$.

Safety

Lasers are potentially **dangerous**. Strict precautions should be observed in their use:

- Never look directly into the beam, or allow anyone else to.
- Never point the beam at anyone.
- Avoid reflections of the beam.
- Always keep your back to the laser when it is on.
- Always switch off the laser when not in use.
- Always work in a well-lit room to prevent the pupils of your eyes dilating.

Worked example

The photograph in Figure 44(a) is *one-quarter* full scale.

a The distance from the grating to the screen was 200 ± 2 mm. Take such measurements as are necessary to determine the angles of diffraction for the first three orders of the green light. Record your measurements in a suitable table.

b The manufacturer states that the grating has a spacing of 300 ± 5 lines per mm. What value does this give for the line separation d?

c Use this, with your values for θ, to find an average value for the wavelength of the green light.

d Estimate the uncertainty in your value for the wavelength.

e Discuss two ways in which you could reduce this uncertainty.

Answer

a

Order	Maxima separation/mm	x/mm	$\tan\theta$	θ/°
1	17.0	34.0	0.170	9.7
2	34.5	69.0	0.345	19.0
3	53.0	106.0	0.530	27.9

b Line separation $d = 1/(300 \times 10^3)$ m $= 3.33 \times 10^{-6}$ m

Percentage uncertainty in manufacturer's value is $(5/300) \times 100\% = 1.7\%$

Uncertainty in $d = 1.7\%$ of 3.33×10^{-6} m $= \pm 0.07 \times 10^{-6}$ m

The value for d should therefore be stated as $(3.33 \pm 0.07) \times 10^{-6}$ m

Exam tip

You should always try to make measurements that are as large as possible to minimise experimental uncertainty.

Exam tip

You must be able both to draw, and to use, scaled diagrams.

Exam tip

Try to interpolate between scale divisions, e.g. to 0.5 mm on a millimetre scale, as here.

c From $n\lambda = d\sin\theta$ we obtain $\lambda = \dfrac{d\sin\theta}{n}$

$n = 1$: $\lambda = \dfrac{3.33\times10^{-6}\,\text{m}\times\sin9.7°}{1} = 564\,\text{nm}$

$n = 2$: $\lambda = \dfrac{3.33\times10^{-6}\,\text{m}\times\sin19.0°}{2} = 544\,\text{nm}$

$n = 3$: $\lambda = \dfrac{3.33\times10^{-6}\,\text{m}\times\sin27.9°}{3} = 520\,\text{nm}$

Average $\lambda = 543\,\text{nm}$

d The range of values is $(564 - 520)\,\text{nm}$, so uncertainty = ½ range = $\pm22\,\text{nm}$

% uc = $(22\,\text{nm}/543\,\text{nm}) \times 100\% = 4.0\%$

% uc in d = 1.7% (see **b** above)

% uc in screen distance = $(2\,\text{mm}/200\,\text{mm}) \times 100\% = 1.0\%$

% uc in λ = 4.0% + 1.7% + 1.0% = 6.7%

Uncertainty in λ = 6.7% × 543 nm = 36 nm

We can therefore say that $\lambda = 543 \pm 36\,\text{nm}$

e The uncertainty could be reduced by:

- moving the screen further away (provided the pattern remains visible) so that the distance between the maxima is greater
- using a grating with smaller line separation, e.g. 600 lines per mm, so that the maxima are further apart

Essential theory: Young's double slit experiment

The wavelength of laser light can also be determined by using a double slit to produce an interference pattern — Young's fringes. The apparatus is set up as shown in Figure 46.

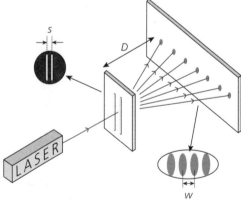

Figure 46

By measuring the slit spacing s, the fringe spacing w and the slits–screen distance D, the wavelength of the light used can be calculated from:

$$w = \frac{\lambda D}{s}$$

Knowledge check 60

Assume that the diffraction grating and screen distance in the worked example are kept the same.

a Explain the difference you would observe in the distances between the maxima if the green laser was replaced by a red laser emitting a wavelength of 693 nm.

b (A-level only) Determine how many orders can be observed with this red laser.

Exam tip

For a question like **e**, which asks you to *explain*, you must both state what you would do *and* explain how this would reduce the uncertainty.

Worked example

Students used a red laser to illuminate a double slit with slit spacing = 0.25 mm. They measured the fringe spacing and calculated the wavelength for six different slits–screen distances, as shown in Table 9.

Table 9

D/m	3.00	2.50	2.00	1.50	1.25	1.00
w/mm	7.5	6.3	5.1	3.8	3.1	2.4
λ/nm	625	630	638	633	620	600

a Describe how they should try to determine the fringe spacing accurately.

b Calculate the mean value for the wavelength, together with the uncertainty in this value.

c The manufacturer states the wavelength of the laser to be 630 nm. Discuss whether or not the students' results are consistent with this value.

d The students' teacher suggests that the wavelength could be found by means of a graph. What graph should be plotted, and how could the wavelength be found from the graph?

e Explain what would be seen if a green laser was used to illuminate the double slits.

Answer

a The students should measure across several fringes. The fringe spacing w can then be obtained by dividing this measurement by the number of fringes. It is sometimes easier to take measurements from the centres of the dark fringes, as the centres of the bright fringes may be more difficult to locate than the centres of the dark fringes.

b Mean wavelength = (625 + 630 + 638 + 633 + 620 + 600) nm/6 = 624 nm
The uncertainty can be estimated by calculating half the range:
½ × (638 nm − 600 nm) = 19 nm $\Rightarrow$ wavelength = 624 ± 19 nm

c As the stated value of 630 nm falls within this range, the students' value is consistent with the manufacturer's value, allowing for the uncertainty of the experiment.

d From the formula $w = \dfrac{\lambda D}{s}$ we can see that a graph of w against D should be plotted. Then the gradient will be $\lambda/s \Rightarrow \lambda$ = gradient × s.

e The fringe spacing is directly proportional to the wavelength of the light used. As green light has a shorter wavelength, the green fringes would be closer together.

Exam tip

Remember to convert all distances into metres before substituting into the equation. Visible light wavelengths are usually quoted in nanometres (1 nm = 10⁻⁹ m).

Knowledge check 61

a Explain what is meant by coherent sources.

b Explain why double slits emit light that is coherent.

c Calculate the wavelength of the light being used if the two slits are 0.50 mm apart, the distance to the screen is measured to be 0.80 m and the fringe spacing is 1.0 mm.

Required practical 3

Determination of g by a free-fall method

Essential theory

A free-falling object means one that is falling vertically under gravity, with no other forces acting. In practice, an object falling in air will experience air resistance. But provided the object is made of a dense material and its speed is not excessive, it may be considered to be falling freely. We can then apply the equations of uniformly accelerated motion to the object — in particular:

$$s = ut + \tfrac{1}{2}at^2$$

If the object is released from rest, $u = 0$. If the object then falls through a height h with acceleration g we get:

$$h = \tfrac{1}{2}gt^2$$

Rearranging gives us:

$$t^2 = \frac{2}{g} \times h$$

If we plot a graph of t^2 on the y-axis against h on the x-axis, the gradient will be $2/g$.

Planning

There are several ways of doing this experiment: using light gates with a computer interface, strobe photography and video-frame analysis are examples. But, in all cases, g is calculated from the time taken for a free-falling object to move through a measured distance. The main problem is the relatively short times involved.

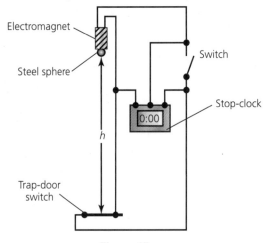

Figure 47

Figure 47 shows one possible method. The points to remember are as follows:
- The time is measured using an electronic stopclock that reads to 1/100th of a second.
- When the switch is in the position shown, the electromagnet is activated and holds the steel ball bearing in place.
- When the switch is thrown, the electromagnet circuit is broken, releasing the ball, and simultaneously the clock circuit is activated, starting the clock.
- The clock is stopped when the ball hits the trapdoor and breaks the circuit.
- The time should be determined at least three times and averaged for each height, over as wide a range of heights as possible.

Safety

The following precautions should be observed:
- Although low voltages are being used, care should be taken with wiring to prevent any short circuits.
- A suitable container should be placed under the trapdoor to catch the ball bearing to prevent it falling on the floor and creating a trip hazard.

Knowledge check 62

Show that the time taken for an object to fall 1.2 m under gravity is about 0.5 s.

Worked example

Table 10 shows a typical set of data. In order to simplify this example, the times indicated are the average of three values for each height. In practice *all* the values should be shown.

Table 10

h/m	t/s	t²/s²
0.600	0.38	
0.800	0.44	
1.000	0.47	
1.200	0.52	
1.400	0.56	
1.600	0.59	

a Complete the table to show the values of t^2.

b Plot a graph of t^2 against h.

c Determine a value for g from the gradient of your graph.

d Estimate the uncertainty in your value for g, with an explanation of your reasoning.

e Suggest why the graph has a small positive intercept on the t^2 axis.

Answer

a The values of t^2/s^2 are: 0.14, 0.19, 0.22, 0.27, 0.31 and 0.35, respectively.

b Your graph should be like Figure 48:

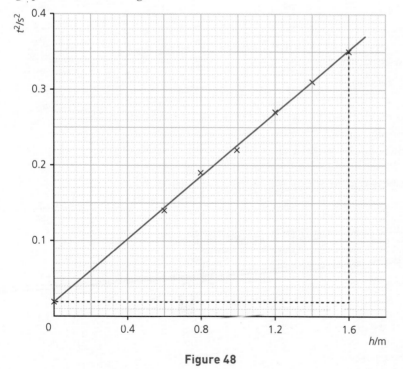

Figure 48

c $\quad \text{gradient} = \dfrac{(0.35-0.02)\,\text{s}^2}{(1.60-0.00)\,\text{m}} = 0.20(6)\,\text{s}^2\,\text{m}^{-1} = \dfrac{2}{g}$

This gives $g = \dfrac{2}{0.20(6)\,\text{s}^2\text{m}^{-1}} = 9.7\,\text{m}\,\text{s}^{-2}$

d As the values of t could only be measured to 2 sf, it would be reasonable to state the value for g as $9.7 \pm 0.1\,\text{m}\,\text{s}^{-2}$. Without all the timings, a more detailed analysis is not possible.

e A small positive intercept on the t^2 axis suggests a systematic error, either in t (e.g. a delay in releasing the ball after the switch has been thrown or a delay in the trapdoor opening after the ball has hit it) or in the measurement of h (all the h values being too small).

Exam tip

Remember to draw a large triangle when determining a gradient and to show the coordinates you have taken in your calculation.

Required practical 4

Determining the Young modulus of a material by a simple method

Essential theory

The Young modulus E of a material is defined by the equation:
$$E = \frac{\text{stress}}{\text{strain}} = \frac{F/A}{\Delta l/l} = \frac{Fl}{A\Delta l}$$
Rearranging, we get:
$$F = \frac{EA}{l}\Delta l$$

A graph of the applied force F on the y-axis against the extension Δl on the x-axis will have a gradient EA/l. If we measure the length l, and determine A from a measurement of the diameter d, we can then determine E.

Knowledge check 63

Show that a 3 m length of 28 swg copper wire (diameter 0.376 mm) will extend by about 1 mm when a force of 5 N is applied. The Young modulus of copper is $1.2 \times 10^{11}\,\text{Pa}$.

Planning

The extension of even a thin copper wire is only in the order of millimetres, which makes its measurement tricky. Although special apparatus, incorporating a vernier or micrometer, can be used, a reasonable value can be obtained for the Young modulus by the simple apparatus shown in Figure 49. This has the added advantage that the behaviour of the wire can be investigated up to when it breaks — but see 'Safety' below.

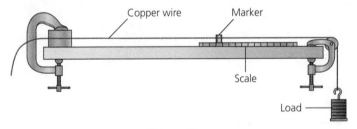

Figure 49

The following procedure should be carried out:

■ The thin (e.g. 26 swg or 28 swg) copper wire should be tightly clamped to avoid slipping and should be as long as possible — usually about 3 m.

- Weights are added until the wire is just taut and the initial reading of the marker is taken.
- The extension can then be determined for additional loads.

Safety

The following precautions should be observed:

- Place a suitable 'crash pad' under the weights to prevent them hitting the floor directly when the wire breaks.
- Do not put your feet under the weights.
- Wear goggles for protection when the wire breaks.

Worked example

Table 11 shows a typical set of measurements. As the extensions are small, an attempt has been made to record the extensions to a precision of 0.5 mm.

Table 11

Mass added/kg	F/N	Δl/mm	Diameter d/mm
0.50	4.9	1.5	0.38, 0.37, 0.39, 0.38,
1.00	9.8	2.5	0.39, 0.37
1.50	14.7	3.0	Average d = 0.38 ± 0.01 mm
2.00	19.6	4.0	
2.50	24.5	5.5	Length l/m
3.00	29.4	6.0	2.95, 2.93
3.50	34.3	7.5	Average l = 2.94 ± 0.01 m
4.00	39.2	10.0	
5.00	Large extension before wire breaks		

a Explain how the values for the diameter and the length of wire shown in the table could be determined.

b Plot a graph of load on the y-axis against extension on the x-axis, using error bars to reflect the precision of ±0.5 mm in the values of extension (although the extension is the dependent variable, it is conventional to plot load against extension or stress against strain).

c Use the gradient of your graph and the data in the table to determine a value for the Young modulus for copper.

d Estimate the uncertainty in your value for the Young modulus.

Answer

a The length can be found using three metre rules (or a 3 m tape measure if available). Even if the precision is only 1 cm, this gives rise to an uncertainty of less than 1%. The length should be measured *from the point where the wire is clamped to the reference point on the marker.*

The diameter should be found in *different places and at different orientations* along the length of the wire using a digital vernier or micrometer having a precision of 0.01 mm or better.

b Your graph should look like Figure 50.

Exam tip

It is good practice to try to interpolate between the scale readings of an instrument if possible, e.g. to 0.5 mm on a millimetre scale or to 0.1°C on a thermometer having graduations of 1°C.

Exam tip

Note that to get full marks the phrases in italics *must* be included.

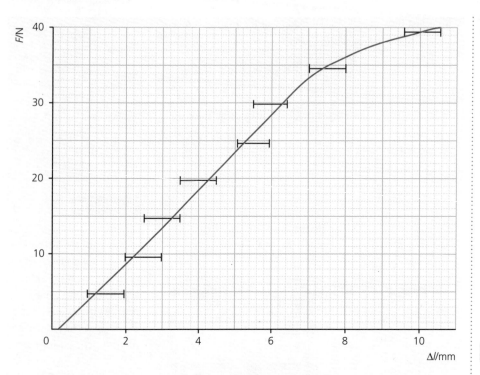

Figure 50

c Determine the gradients for lines of maximum and minimum slope that fall within the error bars. Use the average of these values for your gradient value. This should be approximately $4800 \pm 400\,\text{N}\,\text{m}^{-1}$.

$$E = \text{gradient} \times \frac{l}{A} = 4800\text{N}\text{m}^{-1} \times \frac{2.94\text{m}}{\pi\left(\dfrac{0.38\times10^{-3}\text{m}}{2}\right)^2}$$

$$= 1.2(4)\times10^{11}\text{Pa}$$

d % uc in diameter $= \dfrac{0.01\text{mm}}{0.38\text{mm}}\times100\% = 2.6\%$

% uc in area $A = 2 \times 2.6\% = 5.2\%$

% uc in $l = \dfrac{0.01\text{m}}{2.94\text{m}}\times100\% = 0.3\%$

% uc in gradient $= \dfrac{400\text{N}\text{m}^{-1}}{4800\text{N}\text{m}^{-1}}\times100\% = 8.3\%$

% uc in $E = (5.2 + 0.3 + 8.3)\% = 13.8\%$

Uncertainty in $E = 13.8\%$ of $1.2(4) \times 10^{11}\,\text{Pa} = \pm0.17 \times 10^{11}\,\text{Pa}$

This gives a value for E of $(1.24 \pm 0.17) \times 10^{11}\,\text{Pa}$ or more realistically $(1.2 \pm 0.2) \times 10^{11}\,\text{Pa}$

Knowledge check 64

Calculate the stress and strain in the wire when it is loaded with a mass of 2.00 kg. Use your values to calculate a value for the Young modulus.

Knowledge check 65

Show that the units in the equation for E are equivalent to Pa.

Exam tip

Remember that, when a quantity is squared, its percentage uncertainty is doubled. In a product, the percentage uncertainty is the sum of the percentage uncertainties of the individual terms. And in a quotient, the percentage uncertainty is the sum of the percentage uncertainties of the numerator and denominator.

Required practical 5

Determination of resistivity of a wire using a micrometer, an ammeter and a voltmeter

Essential theory

The resistivity ρ of a material is given by the equation:

$$R = \frac{\rho l}{A}$$

where R is the resistance of a length l of the material, having area of cross-section A.

If we plot a graph of R on the y-axis against l on the x-axis, we should get a straight line through the origin of gradient ρ/A.

If we determine A for the sample, we can then find ρ from $\rho =$ gradient $\times A$.

Planning

The easiest method is to find the resistance of different lengths of resistance wire, such as nichrome, and measure the diameter d of the wire to determine A from $A = \pi d^2/4$.

Note that:

- As the resistance of a wire such as this is only a few ohms, an ohmmeter is unsuitable for measuring the resistance with any degree of accuracy.
- Therefore, an ammeter and voltmeter must be used, and the resistance determined from $R = V/I$.
- The circuit should be set up as shown in Figure 51(a).
- The movable crocodile clip should be pressed firmly on to the wire to minimise contact resistance (but not too firmly or the wire will be damaged).

(a)
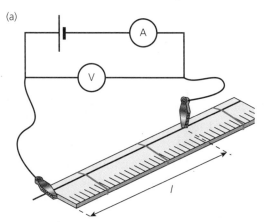

(b)
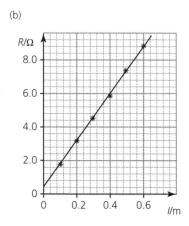

Figure 51

Safety

The following precautions should be observed:

- Although low voltages are being used, care should be taken with wiring to prevent any short circuits.
- See also question **h** in the worked example below.

Knowledge check 66

Show that the resistance of a 50 cm length of 30 swg (diameter 0.3150 mm) nichrome wire is about 7 Ω. The resistivity of nichrome is 1.08×10^{-6} Ω m.

Exam tip

Remember that the unit for resistivity is ohm × metre, i.e. Ω m.

Worked example

a State what instrument you would use to measure the diameter d of the wire. Justify your choice of instrument.

b The following readings were recorded for the diameter of the wire:
d/mm = 0.31, 0.33, 0.30, 0.39, 0.32
Explain what value you would take for the average diameter.

c Estimate the percentage uncertainty in your value for the diameter of the wire.

d Figure 51(b) shows how the resistance R varies with the length l for such a wire. The intercept on the resistance axis indicates a systematic error. Suggest a possible reason for this.

e Determine the gradient of the graph and hence find a value for the resistivity of nichrome.

f Assuming that the other experimental uncertainties are much less than the uncertainty you calculated in c, estimate the uncertainty in your value for the resistivity.

g The cell has an emf of 1.5 V. Show that the current in the wire is about 0.5 A when a length of 20 cm is connected into the circuit.

h Explain the problem that could arise if short lengths of wire are connected. Suggest any additions that you could make to the circuit to overcome this problem.

Answer

a A micrometer (or digital vernier) would be suitable, as it could measure a diameter of about 0.3 mm with adequate precision (0.01 mm or possibly 0.001 mm, depending on the instrument).

b The 0.39 mm value is obviously a spurious reading and so should be ignored when calculating the average. The average is therefore:

$$d = (0.31 + 0.33 + 0.30 + 0.32)\,mm \div 4 = 0.31(5)\,mm$$

c The uncertainty in d should be taken as half the range of values, which is 0.015 mm. The percentage uncertainty in d is therefore:

$$\%\,uc\,in\,d = \frac{0.015mm}{0.315mm} \times 100\% = 5\%$$

As the area is $\pi d^2/4$, the percentage uncertainty in A is 2 × 5% = 10%.

d The intercept indicates a systematic error — a small resistance for zero length of wire. This is probably caused by a poor contact between the crocodile clip and the wire at the zero end, giving rise to a small contact resistance. Although there will also be contact resistance between the movable clip and the wire, this will depend on how hard the clip is pressed against the wire, and so is more likely to give rise to random errors.

e $gradient = \dfrac{(8.80 - 0.40)\,\Omega}{(0.60 - 0.00)\,m} = 14.0\,\Omega\,m^{-1}$
$A = \pi d^2/4 = [\pi \times (0.315 \times 10^{-3}\,m)^2] \div 4 = 7.7(9) \times 10^{-8}\,m^2$
$\rho = gradient \times A = 14.0\,\Omega\,m^{-1} \times 7.7(9) \times 10^{-8}\,m^2 = 1.09 \times 10^{-6}\,\Omega\,m$

f Percentage uncertainty in ρ = percentage uncertainty in A = 10%
Uncertainty in ρ = 10% of $1.09 \times 10^{-6}\,\Omega\,m$ = $0.109 \times 10^{-6}\,\Omega\,m$
The value of ρ is therefore best expressed as $(1.09 \pm 0.11) \times 10^{-6}\,\Omega\,m$ →

Exam tip

The diameter of a wire should be measured in different places along the length of the wire and at different orientations to check its uniformity and get a better average value.

Exam tip

Remember that, if a quantity is squared, then the percentage uncertainty is doubled.

Knowledge check 67

Calculate the diameter of 32 swg gauge nichrome wire, which has a resistance per unit length of $18.3\,\Omega\,m^{-1}$.

[Note that, as there are several values for the diameter, enabling us to get an average with an uncertainty calculated from the spread of readings, we are justified (just) in giving the value for ρ to 3 sf. With a single reading for the diameter, only 2 sf could be justified.]

g The resistance of a 20 cm length of wire is $0.2 \times 14.0\,\Omega = 2.80\,\Omega$

$$I = \frac{V}{R} = \frac{1.5\,\text{V}}{2.80\,\Omega} = 0.54\,\text{A} \approx 0.5\,\text{A}$$

h A current of the order of 0.5 A could heat the wire and therefore change its resistivity. This could be overcome by adding a resistor (e.g. $10\,\Omega$) in series with the cell and ammeter.

Required practical 6

Investigation of the emf and internal resistance of electric cells and batteries by measuring the variation of the terminal pd of the cell with current in it

Essential theory

When there is a current I in an electric cell, the terminal pd V is less than the emf ε of the cell because work has to be done overcoming the internal resistance r of the cell. This can be expressed by the equation:

$$V = \varepsilon - Ir$$

or rearranging:

$$V = -rI + \varepsilon$$

If we measure the pd for different values of current in the cell, a graph of V on the y-axis against I on the x-axis will be a straight line of gradient $-r$ and intercept ε (provided the internal resistance is constant).

The simplest experiment is to do this for a dry cell, such as a 1.5 V AA cell. In the examination, questions may be set using the principles with which you should be familiar, but in a different context. An interesting alternative might be to investigate what happens with a solar cell.

Planning

The circuit required is shown in Figure 52.
- The solar cell is illuminated with a bench lamp (not shown in Figure 52).
- The current in the cell is adjusted with the variable resistor and the corresponding pd recorded.

Safety

The following precautions should be observed:
- Although low voltages are being used, care should be taken with wiring to prevent any short circuits.
- Digital meters should be handled with care to prevent damage.
- Be careful not to touch the lamp — it is likely to be very hot (unless it is an LED).

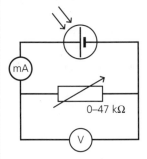

Figure 52

Worked example

A typical set of results is shown in Table 12.

Table 12

I/mA	0.100	0.200	0.300	0.400	0.500	0.600	0.700	0.800	1.000	1.200
V/V	2.63	2.61	2.54	2.52	2.49	2.44	2.38	2.33	2.14	1.70

a A 47-kΩ rotary potentiometer was used for the variable resistor. Use the data in the table to explain why this is a suitable choice.

b What is the control variable that must be kept constant in this experiment?

c Explain why you should keep well away from the cell when taking readings.

d Plot a graph of V/V on the y-axis against I/mA on the x-axis.

e Explain the shape of your graph. Suggest why there is some scatter in the points.

f Use your graph to find the emf of the solar cell and its internal resistance for current values of less than 600 μA.

Answer

a When $I = 1.200$ mA, $V = 1.70$ V, so $R = V/I = 1.70$ V/1.200 mA $= 1.4$ kΩ.

When $I = 0.100$ mA, $V = 2.63$ V, so $R = V/I = 2.63$ V/0.100 mA $= 26$ kΩ.

The resistance needs to have a range from 1.4 kΩ to 26 kΩ. Therefore, a 47 kΩ rotary potentiometer, which can be varied from 0 to 47 kΩ, will be suitable.

b As the voltage from a photocell depends on the illumination, the intensity of the light falling on it is the control variable and must be kept constant. This means that the position of the lamp must not be altered during the course of the experiment.

c You should keep well away from the photocell when taking readings to make sure that you do not get in the way of the light falling on the cell.

d Your graph should look like Figure 53.

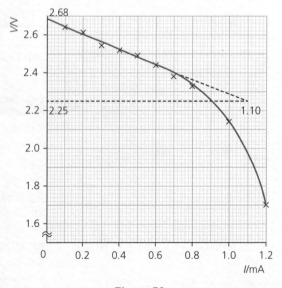

Figure 53

Exam tip

Ideally, the intensity of the light falling on the photocell should be checked during the experiment with a light meter to ensure that it remains constant. The intensity of the light under which the experiment was conducted should then be recorded.

Exam tip

Note that the scale on the y-axis of the graph should *not* start at $V = 0$ or else the scale would be far too small. Your plotted points should occupy *at least* half the space available. Also, the straight part of the line has been extended to give a large triangle for the gradient.

e The graph is initially a straight line of negative slope ($= -r$) indicating a constant internal resistance r up to about $600\,\mu A$. After $600\,\mu A$ the graph curves downwards, indicating that the internal resistance is getting larger (because the pd is decreasing more than it would do if r was constant). The scatter in the points is probably due to slight changes in illumination.

f emf: $\varepsilon =$ intercept on V axis when $I = 0 \Rightarrow \varepsilon = 2.68\,V$

The internal resistance for current values of less than $600\,\mu A$ is given by the gradient of the *linear part* of the graph. Note that this has been extended to give a **large** triangle.

$$r = -\text{gradient} = -\frac{(2.68 - 2.25)\,V}{(0.00 - 1.100)\,mA}$$

$$= 390\,\Omega \text{ to 2sf}$$

Knowledge check 69

When the current in the solar cell is $1.20\,mA$, the pd across its terminals is $1.70\,V$. Show that this gives a value for the internal resistance of about $800\,\Omega$ (see answer **e** in the worked example).

Required practical 7

Investigation into simple harmonic motion using a mass–spring system and a simple pendulum

Essential theory: mass on a spring

You are required to time the vertical oscillations of a mass on a spring and hence find the *natural* frequency of oscillations. You need to recall that a mass m oscillating on a light vertical spring that obeys Hooke's law and has a spring constant k will execute SHM. The time period of the oscillations given by:

$$T = 2\pi\sqrt{\frac{m}{k}}$$

Knowledge check 70

Show that rearranging the equation $T = 2\pi\sqrt{\dfrac{m}{k}}$

gives $T^2 = \dfrac{4\pi^2}{k} \times m$.

Planning

You should note the following:

■ The experiment works well with a standard $20\,mm$, so-called 'expendable', laboratory spring — 'so-called' because your teacher may not be best pleased if you 'expend it'! You should therefore be careful not to stretch it beyond its elastic limit.

■ You also need a range of $100\,g$ masses and a $50\,g$ mass so that the spring can be loaded up to about $400\,g$ in intervals of $50\,g$.

■ A convenient 'unknown' mass could be a block of wood having a volume of about $500\,cm^3$.

■ A stopwatch is needed for timing.

Safety

See question **b** in the worked example.

Worked example

Table 13 shows a typical set of data.

Table 13

m/kg	0.100	0.150	0.200	0.250	0.300	0.350	0.400	Block
T/s	0.410	0.502	0.574	0.648	0.710	0.767	0.819	0.663
T^2/s^2	0.168	0.252						

a Draw a diagram of how you would set up the apparatus.

b State any safety precautions that you would observe in conducting the experiment.

c Suggest why a block of wood having a volume of about $500\,cm^3$ might be a convenient 'unknown' mass.

d Explain the experimental techniques you would use to obtain data such as those in Table 13, adding to your diagram if necessary.

e Discuss why using a graph of T^2 on the y-axis against m on the x-axis is a better way of determining the mass of the wooden block than by plotting a graph of T against m.

f Complete the table and plot a graph of T^2 against m. Use your graph to determine the mass M of the block of wood.

g The block was found to measure $89 \pm 1\,mm \times 75 \pm 1\,mm \times 75 \pm 1\,mm$. Use these data and your mass for the block to determine the density of the wood. Estimate the uncertainty in your value for the density, explaining how you made your estimate.

Answer

a Figure 54 shows how the apparatus should be arranged.

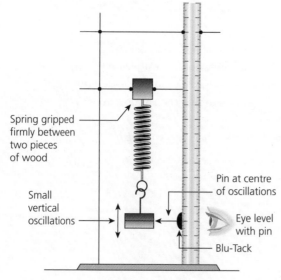

Spring gripped firmly between two pieces of wood

Small vertical oscillations

Pin at centre of oscillations

Eye level with pin

Blu-Tack

Figure 54

b A shock pad should be placed under the masses and the stand should be secured so that it does not topple when the masses are added to the spring.

$\rightarrow$

Knowledge check 71

Use the gradient of your graph to determine a value for the spring constant k of the spring.

c Assuming the density of wood is about $500\,\text{kg m}^{-3}$, a volume of $500\,\text{cm}^3$ $(5 \times 10^{-4}\,\text{m}^3$) would have a mass of about $0.25\,\text{kg}$, which would be in the middle region of the graph.

d The following techniques should be used:
  - Make sure that the spring is gripped firmly so that there is no slipping or twisting.
  - Time, say, 20 *small* oscillations and repeat. Average and divide by 20 to find T.
  - Use a marker at the *centre* of the oscillations to aid counting, as shown in Figure 54.

e A graph of T^2 against m will give a straight line, while a graph of T against m will be a curve. As it is much easier to draw a line of best fit for points that lie on a straight line, a graph of T^2 against m is better.

f The remaining values of T^2/s^2 are 0.329, 0.420, 0.504, 0.588, 0.671 and 0.440. Your graph should be a straight line through the origin, which, by reading off the mass corresponding to $T^2 = 0.440\,\text{s}^2$, gives the mass M of the block as $0.260\,\text{kg}$.

g The volume of the block is $(89 \times 75 \times 75) \times 10^{-9}\,\text{m}^3 = 5.0 \times 10^{-4}\,\text{m}^3$. So

$$\text{density} = \frac{\text{mass}}{\text{volume}} = \frac{0.260\,\text{kg}}{5.0 \times 10^{-4}\,\text{m}^3} = 520\,\text{kg m}^{-3}$$

The percentage uncertainty in the volume is $(1/89 + 1/75 + 1/75) \times 100\% = 3.8\%$. The mass can be read off the graph to about $\pm 0.002\,\text{kg}$, giving a percentage uncertainty of $(0.002/0.260) \times 100\% = 0.8\%$ (or we could say that the timing uncertainty for the block is human error of $0.1\,\text{s}$ in a time of about $13\,\text{s}$ for 20 oscillations of the block — this also gives a percentage uncertainty of about 0.8%). The percentage uncertainty in the density is therefore $3.8\% + 0.8\% = 4.6\%$, giving an uncertainty of:

$$4.6\% \times 520\,\text{kg m}^{-3} = \pm 24\,\text{kg m}^{-3}$$

The density of the wood is probably best expressed as $520 \pm 20\,\text{kg m}^{-3}$ (i.e. somewhere between $500\,\text{kg m}^{-3}$ and $540\,\text{kg m}^{-3}$), as the measurements of the block were only to 2 sf.

Essential theory: simple pendulum

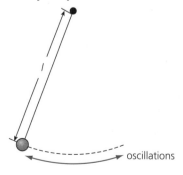

Figure 55

A simple pendulum obeys the defining equation for SHM because the acceleration is proportional to, and in the opposite direction to, the displacement, *provided that the*

Exam tip

A marker ('fiducial' mark) at the *centre* of the oscillations should always be used when timing oscillations — and do not forget to time a *large number* of oscillations and then *divide by the number of oscillations* to find the period. These are points that students often omit and consequently lose marks.

Knowledge check 72

At what point in the motion of a pendulum does it have: **a** maximum potential energy, **b** maximum kinetic energy?

amplitude of the oscillations is small. This leads to the following equation, which can be used to investigate how the time period (T) depends on the length (l):

$$T = 2\pi\sqrt{\frac{l}{g}}$$

Worked example

a What graph should be plotted to confirm the relationship between the period and length for a simple pendulum?

b Explain how the graph in part **a** would be used to determine a value for the acceleration due to gravity.

c Explain why the equation is valid only for small angles.

d Suggest a maximum angle of release for the pendulum during this experiment.

Answer

a T^2 on the y-axis against l on the x-axis would give a straight line through the origin.

b As $T = 2\pi\sqrt{\frac{l}{g}} \Rightarrow T^2 = \frac{4\pi^2}{g} \times l$

The gradient of the graph is therefore $4\pi^2/g \Rightarrow g = 4\pi^2/\text{gradient}$.

c To determine the resultant force acting on the pendulum, the small-angle approximation that $\theta = \sin\theta$ has to be used. Then $F = mg\sin\theta$ becomes $F = mg\theta$, i.e. $F \propto \theta$.

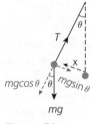

d At an angle of 15°, the difference between $\sin\theta$ (0.259) and θ in radians (0.262) is about 1%. The angle of release should therefore be kept below this.

Figure 56

Required practical 8

Investigating Boyle's (constant temperature) law and Charles's (constant pressure) law for a gas

Essential theory: Boyle's law

You should be familiar with Boyle's law — the pressure of a *fixed mass* of gas is inversely proportional to its volume *provided that the temperature is kept constant.*

Exam tip

When asked to state Boyle's law, you must always give the necessary conditions, namely a fixed mass of gas and a constant temperature. If you do not state the conditions, you will lose marks.

We can express Boyle's law mathematically by:

$$p \propto \frac{1}{V} \Rightarrow p = \text{constant} \times \frac{1}{V} \Rightarrow pV = \text{constant}$$

Knowledge check 73

What length simple pendulum would have a time period of 1.00 s on Earth?

Exam tip

Remember that, in the small-angle approximations $\theta \approx \sin\theta$ and $\theta \approx \tan\theta$, the angle θ must be in radians.

Knowledge check 74

Show that the percentage difference between $\sin\theta$ and θ when $\theta = 20°$ is about 2%.

Knowledge check 75

Sketch graphs that show Boyle's law in the forms p against V and p against $1/V$.

Planning

Two possible ways of doing the experiment are shown in Figure 57.

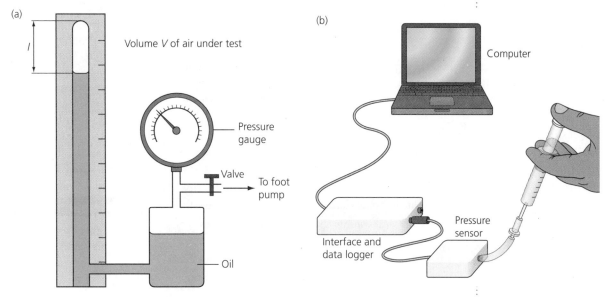

Figure 57

In Figure 57(a), the tube has a uniform bore, so $V \propto l$. In Figure 57(b), the volume of air can be read off directly from the calibrated syringe. In either experiment:

■ as wide a range of values for p and V (or l) as possible should be recorded (but see 'Safety' below)
■ the changes in pressure should be made *slowly*, and time allowed for thermal equilibrium to be reached

Safety

■ It is important to avoid excessive pressure and ensure connections are secure to avoid oil being sprayed at high pressure over observers.
■ The apparatus should be checked by a teacher or technician before use.

Worked example

Table 14 shows a typical set of data.

Table 14

p/kPa	100	120	140	160	180	200	220	240
V/cm³	22.6	18.7	16.1	14.0	12.4	11.1	10.2	9.4
$(1/V)$/10^3 m⁻³	44.2							

a State what are the independent, dependent and control variables in this experiment.
b Explain how the necessary conditions for Boyle's law are ensured.
c How could you, without drawing a graph, quickly check whether the results confirmed Boyle's law? Illustrate your answer with appropriate data.

Exam tip

You need to be confident in unit conversions, e.g. cm³ = (10⁻²m)³ = 10⁻⁶m³.

d Complete the table by adding values of $1/V$. Note that the units are $10^3\,\text{m}^{-3}$. The first value has been done for you.

e Plot a graph of p on the y-axis against $1/V$ on the x-axis and discuss whether it confirms Boyle's law.

Answer

a If *you* change the pressure and then measure the subsequent change in volume, the independent variable is the pressure and the dependent variable is the volume. In either case, the control variables are the mass and temperature of the air.

b To ensure the necessary conditions for Boyle's law:
  ■ the air in the tube or syringe must not be able to leak out so that its mass will remain constant
  ■ the changes in pressure should be made *slowly*, and time allowed for thermal equilibrium to be reached so that the temperature of the air remains constant

c Try multiplying the values of pressure and volume in the table to see if they are equal.
  In kPa cm^3 the values of pV are, to 3 sf: 2260, 2240, 2250, 2240, 2230, 2220, 2260. Within experimental error, these can be considered equal, thereby confirming Boyle's law.

d The remaining values of $1/V$ in $10^3\,\text{m}^{-3}$ are: 53.5, 62.1, 71.4, 80.6, 90.1, 98.0 and 106.4.
  If you plot a graph of p on the y-axis against $1/V$ on the x-axis, you should get a straight line *through the origin*. This shows that $p \propto \dfrac{1}{V}$, which confirms Boyle's law.

Essential theory: Charles's law

Charles's law states that, at constant pressure, the volume of a fixed mass of an ideal gas is directly proportional to its absolute temperature.

We can express Charles's law mathematically by:

$$V \propto T \quad \Rightarrow \quad \frac{V}{T} = \text{constant}$$

If the Kelvin scale is used for temperature, the resulting graph of volume against temperature would be a straight-line graph through the origin (Figure 58).

Planning

A gas syringe and water bath can be used to measure the volume of a gas at different temperatures. Ice cubes can be added to the water to lower the temperature below room temperature to give a greater range.

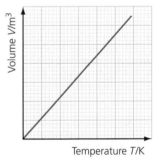

Figure 58

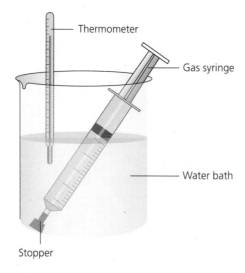

Figure 59

Worked example

A student used the arrangement shown in Figure 59 to measure the volume of air at six different temperatures. A graph of volume V against temperature θ in °C was plotted and extrapolated back to $V = 0\,\text{cm}^3$ in order to determine a value for absolute zero. The student determined absolute zero to be −249°C.

a Sketch a graph of V against θ to show how absolute zero would be determined.
b Calculate the percentage difference between the accepted value of absolute zero and the student's value.
c Give three possible reasons for this difference.

Answer

a

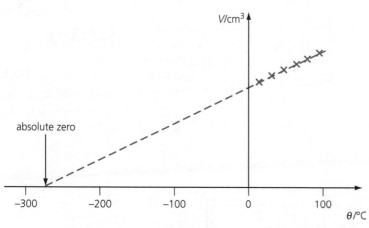

Figure 60

b $[-(273 - 249)°C/(-273°C)] \times 100\% = 8.8\% = 9\%$

c Possible reasons are as follows:

- The air inside the syringe did not have time to come to thermal equilibrium with the water bath, and so the temperature of the trapped air was different from the temperature recorded from the thermometer.
- There is friction between the plunger and the walls of the syringe, which makes it harder for the air inside to expand.
- There could also be a small leak in the stopper, which would mean the mass of trapped air was decreasing.
- The graph requires a long extrapolation, making it difficult to determine an accurate value of the intercept.

Required practical 9

Investigation of the charge and discharge of capacitors

Essential theory

You should be familiar with the formula for a capacitor of capacitance C being discharged through a resistor of resistance R. The potential difference (pd) V across the capacitor decreases exponentially with time t according to the equation:

$$V = V_0 e^{-t/RC}$$

where V_0 is the pd when $t = 0$. The product RC is called the time constant for the discharge. This is the time for the pd to fall to $1/e$ of its initial value, i.e. $0.368V_0$.

You should also be aware that the pd across a capacitor being charged through a resistor increases exponentially with time, although you do not have to know the formula for this. The pd rises to 0.632 of its maximum value in one time constant.

Planning

As capacitors normally have values of the order of microfarads (μF), the time constant will be short unless the resistance is very large — see Knowledge check 78. This means that sometimes it is not possible to investigate the discharge by collecting the data as you go, and so we have to use an oscilloscope, or data logger and computer, to show the charge and discharge.

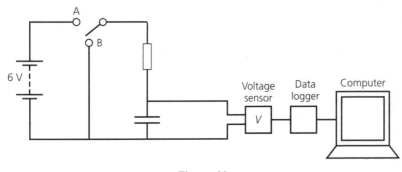

Figure 61

The experimental arrangement is shown in Figure 61. Remember:

- The capacitor is charged through the resistor by connecting the switch to contact A and then discharged through the same resistor by moving the switch to contact B.
- The pd across the capacitor is measured by the voltage sensor and recorded by the data logger.
- The data can then be analysed by the computer.

Safety

- **Danger:** If an electrolytic capacitor is being used, its polarity *must* be observed, i.e. the end marked positive (+) must be connected to the positive side of the power supply.
- Although the pd is low voltage, care should be taken to avoid any short circuits.
- Check that the value of the pd is compatible with the voltage sensor.

Worked example

Figure 62 shows a printout of the charge and discharge of a 22-µF capacitor through a 3.3-kΩ resistor.

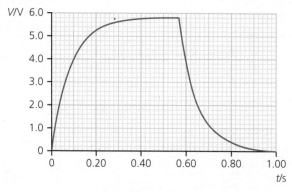

Figure 62

a The value of the resistor is checked with a digital ohmmeter and found to be 3.34 kΩ. Use the discharge curve to determine an experimental value for the capacitance of the capacitor. Estimate the percentage uncertainty in your value.

b The manufacturer's tolerance is ±2% for the resistor and ±10% for the capacitor. Discuss the extent to which your values for the resistance and capacitance are compatible with these data.

c Suggest how you could display the data to improve the accuracy of your value for C.

d Explain why processing the data in the form of a graph of $\ln(V/V)$ against t/s would enable you to determine a more accurate value for C.

e Discuss the advantages of using a data logger and computer rather than an oscilloscope for this experiment.

Answer

a You need to make the following points:
  - The pd will fall from 5.8 V to $(0.368 \times 5.8)\,V = 2.1\,V$ in a time RC.
  - The capacitor starts to discharge at 0.57 s and the pd drops 2.1 V after 0.65 s. →

Exam tip

Remember that the values stamped on resistors and capacitors are what are called 'nominal values' and are subject to a manufacturing tolerance. This is usually 1% or 2% for resistors, and can be as much as 10% or even 20% for capacitors.

Exam tip

You must be able to convert a logarithmic/exponential equation into a linear form by taking natural logarithms on both sides of the equation.

- This gives a time constant of $(0.65 - 0.57)\,s = 0.08\,s$.
- From time constant $= RC = 0.08\,s \Rightarrow C = 0.08\,s/(3.34 \times 10^3\,\Omega) = 2.4 \times 10^{-5}\,F = 24\,\mu F$.
- The small scale suggests it would be reasonable to take the uncertainty in RC as $0.01\,s$. This gives a percentage uncertainty of $(0.01\,s/0.08\,s) \times 100\% = 12.5\%$.
- Assuming the ohmmeter has an uncertainty of $0.01\,k\Omega$, the percentage uncertainty in R will be $(0.01\,k\Omega/3.34\,k\Omega) \times 100\% = 0.3\%$.
- This gives a percentage uncertainty in C of $(12.5 + 0.3)\% = 12.8\% = 13\%$.

(You may have been taught that the pd falls to $V_0/2$ in a time of $0.693RC$. This 'half-life' method is equally acceptable and should yield a similar value for C, bearing in mind the uncertainty in reading off the times.)

b The difference between the stated and measured values for R is:

$(3.34 - 3.3)\,k\Omega = 0.04\,k\Omega$.

The percentage difference is therefore $(0.04\,k\Omega/3.3\,k\Omega) \times 100\% = 1.2\%$, which is well within the 2% tolerance stated.

The difference between the stated and measured values for C is $(24 - 22)\,\mu F$. This is a percentage difference of $(2\,\mu F/22\,\mu F) \times 100\% = 9\%$, which is within the 13% uncertainty calculated for C and within the stated tolerance of 10%.

c The data could be displayed on a larger scale by 'zooming in' on the points where the discharge starts and where the pd is $2.1\,V$ to get more accurate values for the times.

d From $V = V_0 e^{-t/RC} \Rightarrow \ln V = \ln V_0 - \dfrac{t}{RC}$

A graph of $\ln(V/V)$ against t/s would give a straight line of gradient $-(1/RC)$, from which RC can be found. A more accurate value for C should be obtained, as a straight-line graph effectively averages several readings.

e An oscilloscope would not be able to:
- give a printout of the actual data
- process the data
- give data directly to enable a graph of $\ln(V/V)$ against t to be plotted

Knowledge check 79

What is the current in the capacitor in the worked example when it just starts to discharge?

Required practical 10

Investigate how the force on a wire varies with magnetic flux density, current and length of wire using a top pan balance

Essential theory

A current-carrying conductor experiences a force when it is placed inside a magnetic field. This is because the magnetic field due to the current in the wire adds to the field due to the permanent magnets. Magnetic flux density is a vector quantity, and so there are places where the resultant field is stronger and other places where it is weaker. The resultant field pattern is sometimes known as a 'catapult field'.

Exam tip

In a question like this, a diagram is *essential*.

The force F (N) acting on a wire of length l (m) in a magnetic field B (T) at an angle θ to the field (Figure 63) is given by:

$$F = BIl\sin\theta$$

If the wire is at right angles to the field, $\sin\theta = 1$ and so $F = BIl$.

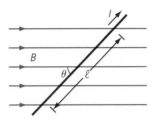

Figure 63

As the forces involved are small ($\sim$ mN), a sensitive electronic balance has to be used to measure the force on the wire. Newton's third law states that the magnetic force experienced by the wire will be equal in size, but in the opposite direction to, the force on the permanent U-shaped 'magnadur' magnets. The change in reading m on the balance is an indication of the magnetic force $F = mg$ on the wire.

Planning

The experimental arrangement is shown in Figure 64. The magnets have their poles on their faces and are fixed to a steel yoke, creating a uniform field. It is important to check that the magnets attract to ensure a north pole faces a south pole. The thick, rigid wire is supported horizontally, perpendicular to the field lines. The procedure is then as follows:

- The balance is zeroed and the current switched on.
- The current and balance reading are recorded.
- The current is increased and readings taken for at least six sets of readings.
- The length of wire inside the field can be doubled by adding an additional yoke and set of magnets.

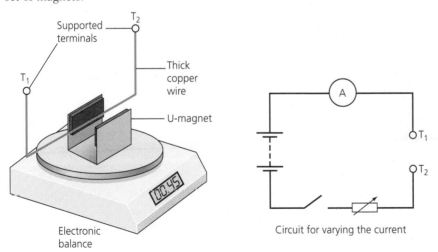

Figure 64

Knowledge check 80

Explain what is meant by Fleming's left-hand rule.

Knowledge check 81

A 45 mm length of wire, in which there is a current of 1.0 A, is at right angles to a magnetic field of strength 0.11 T. Show that the force acting on it is about 0.5 mN.

Exam tip

You should always draw circuit diagrams for any experimental arrangement.

Safety

- In order to produce measurable magnetic forces, relatively large currents have to be used. This causes the wire to heat up, so the circuit should be turned off between readings.
- The rheostat needs to have a large power rating in order to cope with the large current and heat produced.

Worked example

a Explain why you must ensure that the wire is perpendicular to the field lines.

b The data shown in Table 15 were recorded.

Table 15

Current/A	Reading on balance/g	Force/$\times 10^{-3}$ N
0.22	0.08	
0.50	0.17	
1.04	0.37	
1.55	0.54	
2.10	0.74	
2.51	0.88	

Complete the third column, showing the magnetic force on the wire.

c Plot a suitable graph and use it to determine the magnetic flux density. The length of wire in the field was 45 mm.

d Discuss the design flaws of the experiment that reduce the accuracy of the measurement of B.

Answer

a You are investigating the relationship $F = BIl \sin \theta$ where θ = angle between the wire carrying the current and the flux lines. When $\theta = 90°$ the force on the wire will have its maximum value of $F = BIl$.

b

I/A	m/g	F/$\times 10^{-3}$ N
0.22	0.08	0.78
0.50	0.17	1.67
1.04	0.37	3.63
1.55	0.54	5.29
2.10	0.74	7.45
2.51	0.88	8.62

$\rightarrow$

Exam tip

Remember how to convert from a balance reading in grams to a force in newtons.

c

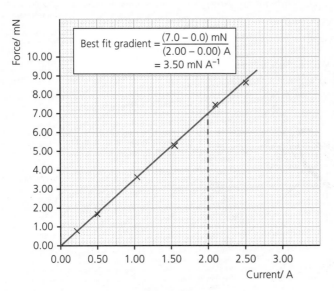

Figure 65

$$\text{gradient} = B \times l \Rightarrow B = \text{gradient}/l = \frac{3.50 \times 10^{-3}\,\text{N\,A}^{-1}}{0.045\,\text{m}} = 0.078\,\text{T} = 78\,\text{mT}$$

d You might make the following points:

■ The magnetic field is not completely uniform — there is an 'edge effect', which makes it weaker near each end.

■ The field also extends slightly beyond each end of the magnets and so the exact length of wire in the field cannot be determined.

■ This could be overcome by using a wire shorter than the length of the magnadur magnets.

■ The connections to the thick wire will also create a field around them and may exert an unknown force on the wire.

■ In light of the above, the experiment can only give an average value for the field between the magnets.

Required practical 11

Investigate, using a search coil and oscilloscope, the effect on magnetic flux density of varying the angle between a search coil and magnetic field direction

Essential theory

A search coil is a small flat coil made from 500–2000 turns of insulated wire and mounted on a handle (Figure 66).

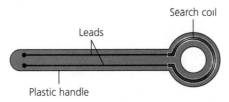

Figure 66

An emf is induced in the coil when it is placed inside a varying magnetic field. If the field is varying at a constant rate, the amplitude of the induced emf will be directly proportional to the amplitude of the magnetic flux density of the field.

For a coil of N turns of area A with its *axis* at an angle θ to a magnetic field of strength B, the flux linkage is calculated using the equation:

flux linkage $= BAN \cos \theta$

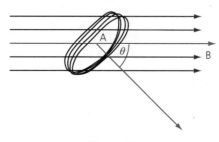

Figure 67

Planning

The experimental arrangement is shown in Figure 68.

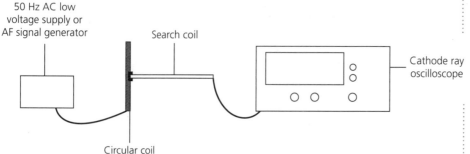

Figure 68

- The search coil is clamped initially with its face perpendicular to the flux lines produced by the circular coil (i.e. its *axis* parallel to the B field, so that $\theta = 0$).
- As $\cos \theta = 1$, this position will produce maximum flux linkage and maximum induced emf in the search coil.
- The time base on the oscilloscope is switched off, so the induced emf is displayed as a vertical line on the screen.
- Measurements are taken of the length of this line.
- The amplitude of the induced emf is then calculated using half the length of this line and by referring to the y-gain setting on the oscilloscope.
- The search coil is rotated to a new angle in the magnetic field and the amplitude of the induced emf is recorded at each angle.
- A suitable means for determining the angle needs to be devised. How you attempt to do this will depend on your particular apparatus and so we cannot be prescriptive here. As you are expected to *demonstrate measurement strategies in order to ensure suitably accurate results*, this is something that you can think about for yourself.

Knowledge check 82

Calculate the flux linking a circular coil of 500 turns and diameter 12.0 mm, where the angle between the field lines and the plane of the coil is 30°. The magnetic flux density = 0.20 T.

Exam tip

Remember that the angle used to calculate flux linkage is the angle between the field lines and the *normal* to the coil's surface (which is the *axis* of the coil).

Knowledge check 83

Explain why an emf is induced in the search coil when there is an alternating current in the large coil.

Exam tip

You should always try to devise ways to make measurements easy to make. This will reduce the uncertainty in your values.

Safety

As the circular coil is taking relatively large currents, it will heat up — do not leave the current on for longer than necessary.

Worked example

A student places a search coil of radius 10mm and 2000 turns in a varying magnetic field of maximum flux density 0.20T with its face perpendicular to the flux lines. The search coil is then connected to an oscilloscope. The induced emf is determined at 15° intervals from the vertical trace on the oscilloscope as the search coil is turned through 360°.

a Calculate the maximum flux linkage.

b Sketch a graph showing how the flux linkage varies with the angle as the coil is rotated. Add scales to the x- and y-axes.

c Explain how the graph would be different if a search coil of radius 20mm was used.

d Use the oscilloscope trace in Figure 69 to determine the amplitude of the induced voltage being displayed.

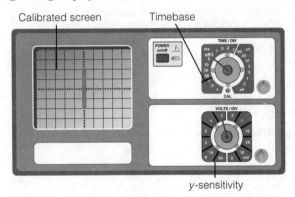

Figure 69

e Sketch the graph that the student should plot to investigate the relationship between flux linkage and the angle between the axis of the search coil and the magnetic flux.

f Explain two strategies the student could use to reduce the uncertainty of the measurements.

Answer

a Flux linkage $= BAN = 0.20\,\text{T} \times \pi(10 \times 10^{-3})^2\,\text{m}^2 \times 2000\ \text{turns} = 0.13\,\text{Wb-turns}$

b

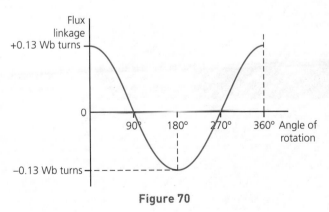

Figure 70

c If the radius is doubled, the area of the coil is increased by a factor of 4. The graph would be the same shape but would show a maximum flux linkage of $4 \times 0.13 = 0.52$ Wb-turns.

d The y-gain is set to 0.5 V/div and the line covers 4.0 divisions in total. Therefore, the peak-to-peak voltage is 4.0 div $\times$ 0.5 V/div = 2.0 V, giving the *peak* emf as 1.0 V.

e Your graph should be labelled with flux linkage on the y-axis and $\cos \theta$ on the x-axis. As the amplitude of the induced emf is proportional to the flux linkage, you should sketch a straight line through the origin.

f Some strategies might be the following:

- As it is difficult to accurately measure the angle the coil makes with the field lines, repeat measurements of induced emf for each angle should be taken, which would enable mean values to be calculated.
- The oscilloscope y-gain should be adjusted to give the largest line that still fits on the screen to increase the precision in the measurement of the induced emf.
- The focus setting on the oscilloscope should be adjusted to give the sharpest line possible, enabling the most accurate measurement of induced emf to be made.
- An ammeter and rheostat should be included in the circular coil circuit to ensure that the current, and therefore the amplitude of the magnetic field, remains constant.

Required practical 12

Investigating the inverse square law for gamma radiation

Essential theory

You should know that gamma radiation is a high-frequency electromagnetic wave and that gamma radiation from a point source spreads out as it travels away from the source. As gamma radiation passes through air without being absorbed, the intensity (measured in $\mathrm{W\,m^{-2}}$) decreases with distance according to an 'inverse square law'. In this practical, your task is to verify the inverse square law for gamma radiation.

Planning

The apparatus is set up as shown in Figure 71. Initially, the cobalt-60 source should not be in the laboratory.

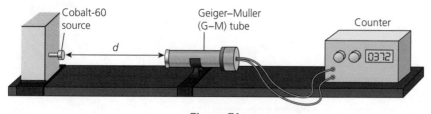

Figure 71

The procedure is then as follows:

- Measure the background count rate, C_0, without the source of gamma radiation.
- Place the source about 1 m from the Geiger tube and measure the distance d between the end of the tube and the front of the source.

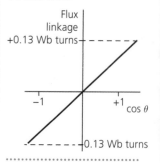

- Measure and record the number of counts in 60 s. Repeat this three times, and then work out the average count rate, C.
- Repeat the measurement for at least six different distances d.

Safety

Danger: radiation can kill. The following safety precautions *must always be adopted* when using radioactive sources:

- Keep as far away as possible (at least 30 cm) from all laboratory sources of ionising radiation.
- Do not touch radioactive materials — use a handling tool.
- Keep sources in their lead storage containers, out of the laboratory, when not in use.
- During an investigation, keep the source pointed away from the body, especially the eyes.
- Limit the time of use of sources — return to secure storage as soon as possible.
- Wash your hands after working with a radioactive source.

Worked example

Table 16 shows a typical set of data. The background count was determined to be 18 counts per minute.

Table 16

d/cm	1.0	2.0	3.0	4.0	5.0	6.0	7.0	8.0
Average count C/min	4310	2600	1950	1380	1100	810	730	610
Corrected count C_c/min	4292							
$1/\sqrt{C_c}$ /min	0.015							

a Describe how you would determine the background count rate.

b Complete the corrected count rate and $1/\sqrt{C_c}$/min rows of the table.

c Plot a graph of d on the y-axis against $1/\sqrt{C_c}$/min and explain whether the data show that the variation of intensity of gamma radiation with distance obeys an inverse square law.

d Use your graph to determine a value for d_0, the intercept when $1/\sqrt{C_c}$/min $= 0$.

e Explain why there is more scatter in the count rate data at greater distances.

Answer

a The background count should be recorded for at least 5 minutes, repeated and averaged. It should be measured with the source outside the laboratory.

b

d/cm	1.0	2.0	3.0	4.0	5.0	6.0	7.0	8.0
C/min	4310	2600	1950	1380	1100	810	730	610
C_c/min	4292	2582	1932	1362	1082	792	712	592
$1/\sqrt{C_c}$ /min	0.015	0.020	0.023	0.027	0.030	0.036	0.037	0.041

$\rightarrow$

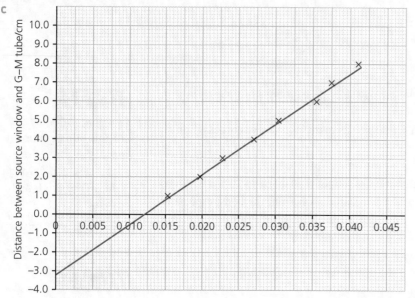

Figure 72

Knowledge check 86

Assuming that gamma radiation follows an inverse square law, calculate the ratio: (intensity of radiation at 4 m) : (intensity of radiation at 1 m).

The graph is a straight line, but it does *not* go through the origin — see Figure 72. However, there is a reason for this. The source itself is a small distance inside the container, and the radiation is detected somewhere inside the tube, not at the window. So the true distance between the source and the place where the radiation is detected is $r = d + d_0$, where d_0 is a correction factor. This is shown in Figure 73.

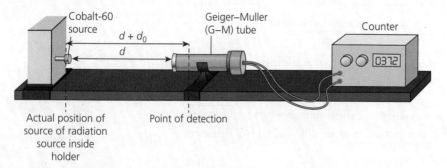

Figure 73

If the intensity, I, of the radiation depends on the distance, r, from the source in accordance with the inverse square law, then I is proportional to $1/r^2$.

As the corrected count rate C_c (i.e. the count rate due to the source) is proportional to the intensity of the radiation, we have:

$$C_c = \frac{k}{(d + d_0)^2}$$

where k is a constant. Rearranging:

$$d = \sqrt{k} \times \frac{1}{\sqrt{C_c}} - d_0$$

which is of the form $y = mx + c$, where gradient $= \sqrt{k}$ and y-intercept $= -d_0$. As Figure 72 is of this form, it indicates that the gamma radiation shows an inverse square law.

d The y-intercept $= -3.2\,\text{cm}$, which is the correction factor.

e At greater distances, the count is lower. The randomness of the decay therefore becomes more apparent.

Questions & Answers

The AQA examinations

The AQA A-level Physics examination consists of three papers. All papers will examine 'Working as a Physicist'. Briefly, this means students:

- working scientifically, developing competence in manipulating quantities and their units, including making estimates
- experiencing a wide variety of practical work, developing practical and investigative skills by planning, carrying out and evaluating experiments and becoming knowledgeable of the ways in which scientific ideas are used
- developing the ability to communicate their knowledge and understanding of physics
- acquiring these skills through examples and applications from the entire course

In particular, paper 3 covers the general and practical principles of physics. It is of 2 hours duration and has 80 marks. Section A consists of questions that assess conceptual and theoretical understanding of experimental methods (indirect practical skills) that will draw on students' experiences of the practical experiments. It also includes data analysis questions.

This student guide covers only practical work, mainly through consideration of the 12 required practicals. Therefore, the questions in this section only reflect the type of question that will be set in paper 3 to test your understanding of experimental methods and are not representative of the paper as a whole. Questions on practical work may also be asked in papers 1 and 2, where knowledge of content is being assessed through a practical context.

In the AQA AS Physics examination, there are just two papers, and questions based on practical work are set mainly in paper 2.

A formulae sheet is provided with each test. Copies may be downloaded from the AQA website.

In addition to the written papers, there is a **Science Practical Endorsement**, for which you will be assessed separately. The Endorsement will not contribute to the overall grade for your A-level qualification, but the result will be recorded on your certificate. In order to gain the Endorsement, you need to provide evidence that you have completed a minimum of 12 identified practical activities, as specified by AQA, that demonstrate your competence to:

- follow written procedures
- apply investigative approaches and methods when using instruments and equipment
- safely use a range of practical equipment and materials
- make and record observations
- research, reference and report

More details are given in the AQA specification, to which you should refer.

Command terms

Examiners use certain **command terms** that require you to respond in a particular way, for example, 'state', 'explain' or 'discuss'. You must be able to distinguish between these terms and understand exactly what each requires you to do. Some commonly used terms are listed below:

- Add/Label — Requires the addition to or labelling of a stimulus material given in the question, for example, adding units to a table or labelling a diagram.
- Assess — Give careful consideration to all the factors or events that apply, and identify which are the most important or relevant. Make a judgement on the importance of something, and come to a conclusion where needed.
- Calculate — Obtain a numerical answer, showing relevant working. If the answer has a unit, this must be included.
- Comment on — Requires the synthesis of a number of variables from data/ information to form a judgement.
- Compare and contrast — Looking for the similarities and differences between two (or more) things. Should not require the drawing of a conclusion. The answer must relate to both (or all) things mentioned in the question. The answer must include at least one similarity and one difference.
- Complete — Requires the completion of a table or diagram.
- Criticise — Inspect a set of data, an experimental plan or a scientific statement and consider the elements. Look at the merits and/or faults of the information presented and back judgements made.
- Deduce — Draw/reach conclusion(s) from the information provided.
- Derive — Combine two or more equations or principles to develop a new equation.
- Describe — Give an account of something. Statements in the response need to be developed, as they are often linked, but do not need to include a justification or reason.
- Determine — The answer must have an element that is quantitative from the stimulus provided, or must show how the answer can be reached quantitatively.
- Devise — Plan or invent a procedure from existing principles/ideas.
- Discuss — Involves the following:
  - Identify the issue/situation/problem/argument that is being assessed within the question.
  - Explore all aspects of the issue/situation/... etc.
  - Investigate the issue/situation/... etc. by reasoning or argument.
- Draw — Produce a diagram either using a ruler or freehand.
- Evaluate — Review information, then bring it together to form a conclusion, drawing on evidence, including strengths, weaknesses, alternative actions, relevant data or information. Come to a supported judgement of a subject's qualities and relation to its context.
- Explain — An explanation requires a justification/exemplification of a point. The answer must contain some element of reasoning/justification, which can include mathematical explanations.
- Give/state/name — All of these command words are really synonyms. They generally all require recall of one or more pieces of information.

Give a reason/reasons — When a statement has been made and the requirement is only to give the reasons why.

- Identify — Usually requires some key information to be selected from a given stimulus/resource.
- Justify — Give evidence to support (either the statement given in the question or an earlier answer).
- Plot — Produce a graph by marking points accurately on a grid from data that are provided and then drawing a line of best fit through these points. A suitable scale and appropriately labelled axes must be included if these are not provided in the question.
- Predict — Give an expected result.
- Show that — Prove that a numerical figure is as stated in the question. The answer must be to at least one more significant figure than the numerical figure in the question.
- Sketch — Produce a freehand drawing. For a graph, this would require a line and labelled axis with important features indicated, but the axes are not scaled.
- State what is meant by — When the meaning of a term is expected but there are different ways of how these can be described.
- Write — When the question asks for an equation.

You should pay particular attention to diagrams, drawing graphs and making calculations. Many candidates lose marks by failing to label diagrams properly, not giving essential data on graphs and, in calculations, by not showing all the working or by omitting units.

The answers that follow should not be treated as model answers — they represent the bare minimum necessary to gain the marks. Some questions have tips on how to answer the question, preceded by 🄴. Comments (denoted by 🄴) may either provide useful tips or indicate where candidates often lose marks. Ticks (✓) are included in the answers to show where a mark has been awarded.

■ Practice questions

Multiple choice

Question 1

A student records the diameter of a wire five times using a micrometer reading to 0.01 mm. The readings are 1.17 mm, 1.19 mm, 1.17 mm, 1.16 mm and 1.18 mm. The student averages these readings, but does not take into account the zero error on the micrometer. The average measurement of the diameter is:

A   precise and accurate   C   accurate but not precise

B   precise but not accurate   D   not accurate and not precise   (1 mark)

Question 2

A student is measuring the force on a current-carrying conductor using the apparatus shown in Figure 1. She measures the length of the copper wire in the magnetic field using a 30 cm rule, which has markings every millimetre.

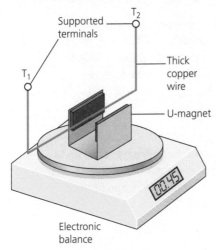

Figure 1

The uncertainty in this measurement of length is:

A ±1 cm **B** ±0.5 cm **C** ±0.2 cm **D** ±0.05 cm (1 mark)

Question 3

In an experiment to determine the specific latent heat of vaporisation of water, steam is condensed in water at room temperature. A student records the initial and final temperatures of the water as $\theta_i = (19.0 \pm 0.5)°C$ and $\theta_f = (29.0 \pm 0.5)°C$. The percentage uncertainty in the temperature difference $\Delta\theta$ is:

A 3% **B** 5% **C** 10% **D** 20% (1 mark)

Question 4

An object falls from rest with an acceleration g. The variation with time t of the displacement s of the object is given by $s = \frac{1}{2}gt^2$. The uncertainty in the value of t is ±3% and the uncertainty in the value of s is ±2%. The best estimate for the uncertainty of the value of g is:

A 5% **B** 6% **C** 8% **D** 11% (1 mark)

Question 5

A student measures the wavelength of a laser using a pair of slits of separation s. Interference fringes are produced on a white screen placed at a distance D from the slits. The separation of the fringes is w.

slits are replaced by new slits of separation $s/2$ and the screen distance is doubled to $2D$. The new fringe separation is:

A $w/2$ **B** w **C** $2w$ **D** $4w$ (1 mark)

Question 6

A student is investigating the absorption of gamma rays by lead. With no absorber, the count rate, when corrected for background, is $400\,min^{-1}$. When a lead disc is placed between the source and detector, the count rate falls to $300\,min^{-1}$. If a second, identical, disc is added the count rate will drop to:

A $150\,min^{-1}$ **B** $200\,min^{-1}$ **C** $225\,min^{-1}$ **D** $250\,min^{-1}$ (1 mark)

Question 7

A student is investigating how the frequency of a stretched wire depends on the mass per unit length μ of the wire using wires of different diameter. The tension and length of each wire are kept constant. The student plots a graph of $\ln f$ on the y-axis against $\ln \mu$ on the x-axis. The gradient will be:

A $-\dfrac{1}{\sqrt{2}}$ **B** $-\dfrac{1}{2}$ **C** $\dfrac{1}{2}$ **D** $\dfrac{1}{\sqrt{2}}$ (1 mark)

Answers to Questions 1–7

1 B

Ⓔ The repeated readings are close to each other, which means that they are *precise*. The student neglected to take into account the zero error and therefore each reading may be larger or smaller than the true reading. The readings are therefore *not accurate*.

2 C

Ⓔ As the measurement of length will be the difference of two readings, each to an uncertainty of 0.1 cm, the uncertainty in the length will be $2 \times 0.1\,cm = 0.2\,cm$. The uncertainty of each reading could be taken as 0.05 cm, giving an overall uncertainty of 0.1 cm, but that is not one of the possible answers.

3 C

Ⓔ When a value is found by adding or subtracting two quantities, the maximum and minimum possible values have to be calculated to find the uncertainty in the value, and then the percentage uncertainty can be calculated.

4 C

Ⓔ Remember to add the percentage uncertainties if quantities are divided; and if a quantity is raised to a power, multiply the percentage uncertainty by the power. Then: % uc in g = % uc in s + $2 \times$ % uc in t.

5 D

Ⓔ Use $w = \lambda D/s$. Practise using ratios, as they are common in multiple-choice questions.

6 C

e With one disc, the count rate reduces to $300\,\text{min}^{-1} = 0.75 \times 400\,\text{min}^{-1}$. So, for two discs, the count rate will decrease to $0.75 \times 300\,\text{min}^{-1} = 225\,\text{min}^{-1}$. This is a quick way of checking whether a graph is exponential.

7 B

e From $\quad f = \dfrac{1}{2l}\sqrt{\dfrac{T}{\mu}} \implies f = \dfrac{\sqrt{T}}{2l} \times \dfrac{1}{\sqrt{\mu}} = \dfrac{\sqrt{T}}{2l} \times \mu^{-1/2} \implies \ln f = \ln\!\left(\dfrac{\sqrt{T}}{2l}\right) - \dfrac{1}{2}\ln\mu$

You need to practise taking logs of equations such as this. Although you could also take logs to base 10 ('log' on your calculator) when trying to find a power, if you *always* take natural logs ('ln'), you will not get the wrong logs when dealing with exponential functions (which *must* be 'ln').

Structured questions

Question 8

A rectangular glass block is measured with a ruler calibrated in 1 mm divisions:

$\qquad$ length = 11.5 cm $\qquad$ width = 6.2 cm $\qquad$ depth = 3.1 cm

(a) Calculate the volume of the block. $\hfill$ (1 mark)

(b) Estimate the percentage uncertainty and hence the absolute uncertainty in your value. $\hfill$ (3 marks)

(c) State how the volume of the block should be recorded. $\hfill$ (1 mark)

$\hfill$ Total: 5 marks

Student answer

(a) volume = 11.5 cm × 6.2 cm × 3.1 cm = 221.03 cm^3 = 221 cm^3 ✓

(b) % uc in volume = % uc in length + % uc in width + % uc in depth

$\quad$ = (0.1 cm/11.5 cm) × 100% + (0.1 cm/6.2 cm) × 100% + (0.1 cm/3.1 cm) × 100% ✓

$\quad$ = (0.870 + 1.61 + 3.23)% = 5.7% ✓

Absolute uncertainty in volume = 5.7% of 221 cm^3 = 13 cm^3 ✓

---

e Remember that the percentage uncertainties are added when quantities are multiplied.

(c) Volume of block = (221 ± 13) cm^3 ✓

---

e Although the width and depth are only measured to 2 sf, an answer of 221 ± 13 cm^3 is arguably better than giving the answer as 220 ± 10 cm^3 to 2 sf. Full credit would be given for either answer.

Question 9

A student is measuring the emf generated by a magnet falling through a coil of wire using a data logger and voltage sensor.

(a) Explain why using a data logger and voltage sensor is a good strategy.　　(3 marks)

(b) After collecting her data, she realises that the sensor has a zero error. Explain what is meant by a zero error, and describe what she should do with her data to take the error into account.

(2 marks)

Total: 5 marks

Student answer

(a) The time for the magnet to fall through the coil is very short. ✓

As a data logger can sample at a high rate, ✓ it can take a large number of readings in this short time. ✓

ℯ As the question says 'explain', it is necessary to state that the time involved is small.

(b) Zero error means that the voltage sensor gives a non-zero reading when there is no measurement being taken. ✓

She should subtract the value of the non-zero reading from her emf data. ✓

Question 10

A student used the arrangement shown in Figure 2 to measure the specific heat capacity of water.

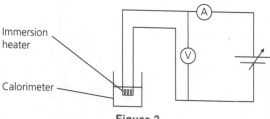

Figure 2

(a) State the measurements the student would need to make in order to measure the specific heat capacity of water.

(2 marks)

(b) Explain how the student could try to make the measurements as accurate as possible.

(4 marks)

(c) State *one* safety precaution you would take if you were to carry out this experiment.

(1 mark)

Total: 7 marks

Student answer

(a) You would need to calculate the energy supplied electrically by measuring the current, pd and time of heating ✓ (or electrical energy supplied $W = VIt$ ✓).

You would also need to measure the mass and the initial and final temperatures of the water. ✓

(b) Points might include the following:

- Stir the liquid regularly to ensure that it is all at the same temperature.
- Record the final steady temperature after switching off to allow equilibrium to be attained.
- Determine the amount of energy absorbed by the container and deduct this from the energy supplied by the heater. This would give you the actual energy supplied to the liquid.
- Insulate the container to prevent energy being lost to the surroundings.
- Leave the thermometer in the liquid to ensure it is at thermal equilibrium with the water and ensure it is not touching the walls of the container.

Any two points ✓✓ explained ✓✓

---

e As the question asks you to 'explain', you must both state the technique *and* explain why it improves the accuracy. Note that using an expanded polystyrene beaker both provides insulation *and* avoids having to determine any energy taken by the container.

(c) Safety precautions could include:

- Make sure the wiring does not short circuit.
- Stand up at all times and clamp the beaker to avoid scalding in case the beaker of hot water is knocked over.
- Keep immersion heater fully immersed.

Any one ✓

Question 11

Figure 3 shows an oscilloscope trace of a sound wave. The signal has a frequency of 2.5 kHz and a peak voltage of 500 mV.

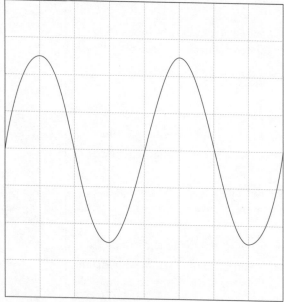

Figure 3

(a) What are the time base and voltage settings on the oscilloscope? (3 marks)

ⓔ You are expected to be familiar with the use of an oscilloscope.

(b) The speed of sound is 330 m s^{-1}. What is the wavelength of this sound wave? (1 mark)

(c) (A-level only) Determine the root mean square voltage displayed on the oscilloscope. (2 marks)

Total: 6 marks

> **Student answer**
>
> **(a)** $T = 1/f = 1/(2500\,\text{s}^{-1}) = 4.00 \times 10^{-4}\,\text{s}$ ✓ $= 400\,\mu\text{s}$
>
> T = time for one wavelength = four divisions $\Rightarrow$ time base = 100 μs/div ✓
>
> $V_{peak} = 500\,\text{mV} = 2.5$ divisions $\Rightarrow$ voltage setting = 500 mV/2.5 div = 200 mV/div ✓

ⓔ To get full marks, you would be expected to express the time base in realistic units — either (preferably) μs/div or ms/div and not 10^{-4} s/div.

> **(b)** $c = f\lambda \Rightarrow \lambda = c/f = 330\,\text{m s}^{-1}/2500\,\text{s}^{-1} = 0.132\,\text{m}$ (or 13 cm) ✓
>
> **(c)** $V_{rms} = \dfrac{V_{Peak}}{\sqrt{2}}$ ✓ $\Rightarrow V_{rms} = \dfrac{500\,\text{mV}}{\sqrt{2}} = 350\,\text{mV}$ ✓

Question 12

A student is investigating the resistance R of conductive modelling putty, which was shaped into a cylinder of length l. She suggests that the relationship between the resistivity ρ of the putty with volume V is given by: $\rho = \dfrac{RV}{l^2}$

The student decides to plot a graph with R on the y-axis and l^2 on the x-axis.

(a) Explain why this is a sensible decision.   (1 mark)

(b) The student obtained the graph shown in Figure 4.

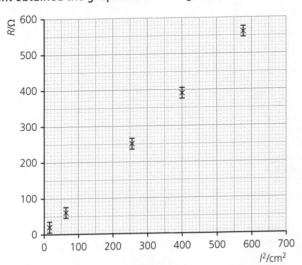

Figure 4

(i) Draw a line of best fit and determine the gradient of this line.

(ii) Determine the uncertainty in your value of gradient. Use an appropriate number of significant figures in your answer.   (6 marks)

(c) The student used $26.8\,\text{cm}^3$ of putty in the experiment. Calculate the resistivity of the putty.   (3 marks)

Total: 10 marks

Student answer

(a) This will produce a straight-line graph with gradient ρ/V. ✓

(b) (i)   gradient $= \dfrac{(600 - 0)\,\Omega}{(620 - 0)\,\text{cm}^2} = 0.97\,\Omega\ \text{cm}^{-2}$ ✓✓

ⓔ Note that a mark is given for drawing a *large* triangle.

(ii) Draw two lines as shown, and calculate their gradients. This will provide the maximum and minimum likely gradient. ✓✓

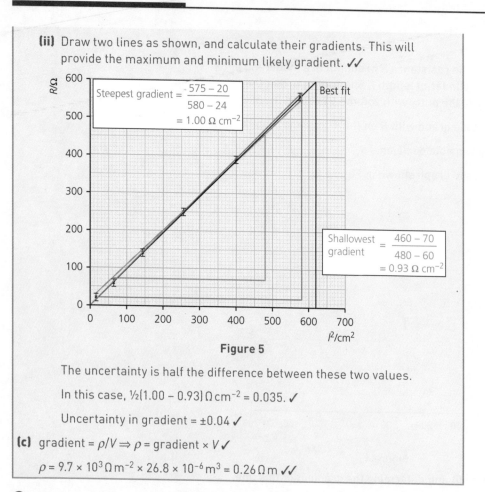

Figure 5

The uncertainty is half the difference between these two values.

In this case, $\frac{1}{2}(1.00 - 0.93)\,\Omega\,cm^{-2} = 0.035$. ✓

Uncertainty in gradient = ±0.04 ✓

(c) gradient = $\rho/V \Rightarrow \rho =$ gradient × V ✓

$\rho = 9.7 \times 10^3\,\Omega\,m^{-2} \times 26.8 \times 10^{-6}\,m^3 = 0.26\,\Omega\,m$ ✓✓

e Watch the units. It is probably best to convert the gradient and volume to metres before doing the final calculation.

Question 13

A student is investigating the properties of a spring. He finds that a mass of 300 g suspended from the spring causes it to extend by 120 mm. When he pulls the mass down a small distance and releases it, he times 20 oscillations in 14.12 s using a digital stopwatch.

(a) Use these data to determine *two* values for the spring constant k of the spring.   (4 marks)

(b) Show that the unit $N\,m^{-1}$ is equivalent to $kg\,s^{-2}$.   (1 mark)

(c) Calculate the percentage difference between your two values and discuss which value is likely to have the greater uncertainty.   (4 marks)

(d) Explain why it is important to pull the mass down only a small distance.   (2 marks)

Total: 11 marks

Student answer

(a) From $F = kx \Rightarrow k = F/x = (0.300\,\text{kg} \times 9.8\,\text{N kg}^{-1})/0.120\,\text{m}$ ✓ $= 24.5\,\text{N m}^{-1}$ ✓

From $T = 2c\sqrt{\dfrac{m}{k}} \Rightarrow k = \dfrac{4\pi^2 m}{T^2} = \dfrac{4\pi^2 \times 0.300\,\text{kg}}{(0.706\,\text{s})^2}$ ✓ $= 23.8\,\text{kg s}^{-2}$ ✓

ⓔ Marks are often lost by either not squaring π or by forgetting to divide the time by the number of oscillations to find T.

(b) $\text{N m}^{-1} = (\text{kg m s}^{-2}) \times \text{m}^{-1} = \text{kg s}^{-2}$ ✓

(c) percentage difference $= \dfrac{24.5 - 23.8}{24.2} \times 100\% = 3\%$ ✓

The uncertainty in measuring the extension is likely to be 2 mm at the most, giving a percentage uncertainty of $(2\,\text{mm}/120\,\text{mm}) \times 100\% = 1.7\%$. ✓

Although the time is recorded to 0.01 s, in reality, the uncertainty is governed by human reaction time, say 0.1 s, giving a percentage uncertainty in T of $(0.1\,\text{s}/14.12\,\text{s}) \times 100\% = 0.7\%$. ✓

The percentage uncertainty in T^2 will therefore be $2 \times 0.7\% = 1.4\%$. ✓

It would therefore appear that the two methods are comparable in terms of uncertainty.

ⓔ It is important that you can make sensible estimations of uncertainties. A common mistake here would be to quote the uncertainty in the timing as 0.01 s, when reaction time is likely to be 0.1 s. Remember, the percentage uncertainty is *doubled* if the quantity is *squared* and the *average* value should be used in the denominator when determining the percentage difference between two *experimental* values.

(d) For SHM, the acceleration and, therefore, the force must be proportional to the displacement. ✓ The spring should be given a *small* displacement to ensure that it oscillates within the Hooke's law region, so that force $\propto$ displacement. ✓

Question 14

Figure 5 shows an arrangement that a student sets up for investigating the acceleration of a glider on a tilted air track. For clarity, the lamps illuminating the light gates have been omitted.

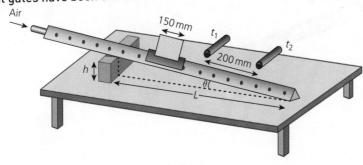

Figure 5

Questions & Answers

(a) The glider is released from rest and the interrupter card, of length 150 mm, is timed as it passes through the two light gates. Times of $t_1 = 1.49$ s and $t_2 = 0.24$ s are recorded.

   (i) Calculate the average velocities v_1 and v_2 of the glider as it passes through each light gate.

   (ii) The light gates are 200 mm apart. Calculate the acceleration of the glider.

ⓔ Using $v^2 = u^2 + 2as$ instead of $v = u + at$ is an alternative way of determining an acceleration when investigating motion.

   (iii) Show that the theoretical acceleration of the glider is $g\sin\theta$, where θ is the angle that the track makes with the horizontal.

   (iv) The student measures the angle by means of a set-square. She records a value of $6.0° \pm 0.5°$. Determine a value for g from the data recorded by the student.

   (v) Estimate the uncertainty in this value for g. What conclusion can you come to regarding the validity of this experiment?

   (vi) Suggest why the value of acceleration obtained is likely to be too small.   (10 marks)

(b) The student's teacher suggests that a much better way to determine θ would be by measuring the distances h and L as shown in Figure 5. The student records $h = 98 \pm 1$ mm and $L = 1000 \pm 1$ mm. What value does this give for θ? Comment on this value.

   (4 marks)

(c) Explain how you could extend this experiment to get a more reliable value for g. Include a sketch of the graph you would plot and how you would find g from it. Explain how your graph would reduce both systematic and random errors.   (6 marks)

Total: 20 marks

Student answer

(a) (i) $v_1 = 0.150 \,\text{m}/1.49 \,\text{s} = 0.101 \,\text{m s}^{-1}$ and $v_2 = 0.150 \,\text{m}/0.24 \,\text{s} = 0.625 \,\text{m s}^{-1}$ ✓

   (ii) From $v^2 = u^2 + 2as \Rightarrow a = \dfrac{v_2^2 - v_1^2}{2s} \Rightarrow a = \dfrac{(0.625\,\text{ms}^{-1})^2 - (0.101\,\text{ms}^{-1})^2}{2 \times 0.200\,\text{m}}$ ✓

   $a = 0.95(1)\,\text{m s}^{-2}$ ✓

   (iii) Refer to the free-body diagram for the glider below.

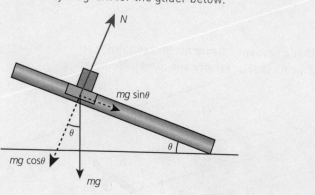

Force on glider F = component of mg down the plane = $mg \sin \theta$. ✓

From $F = ma \Rightarrow a = F/m = (mg \sin \theta)/m = g \sin \theta$. ✓

(iv) From $a = g \sin \theta \Rightarrow g = a/\sin \theta = 0.951\,\mathrm{m\,s^{-2}}/\sin 6.0° = 9.1\,\mathrm{m\,s^{-2}}$ ✓

(v) $0.951\,\mathrm{m\,s^{-2}}/\sin 5.5° = 9.9\,\mathrm{m\,s^{-2}}$ and $0.951\,\mathrm{m\,s^{-2}}/\sin 6.5° = 8.4\,\mathrm{m\,s^{-1}}$ ✓

Uncertainty in $g = \pm 0.8\,\mathrm{m\,s^{-2}}$ ✓

The accepted value of $g = 9.8\,\mathrm{m\,s^{-2}}$ just falls within this (large) uncertainty, so the value of g obtained suggests the experiment could be valid. ✓

(vi) The value obtained is likely to be too small, as there is almost certain to be some frictional force slowing down the glider. ✓

(b) $\tan \theta = h/L = 98\,\mathrm{mm}/1000\,\mathrm{mm} = 0.098 \Rightarrow \theta = \tan^{-1}(0.098) = 5.60°$ ✓

Largest value could be $\tan^{-1}(0.099) = 5.65°$ and smallest value $\tan^{-1}(0.097) = 5.54°$ ✓

So we can say $\theta = 5.60° \pm 0.05°$, ✓ which is a far (10 times) greater precision than using a protractor. ✓

ⓔ Choosing the right instrument, or technique, for making a measurement, and understanding its limitations, are important skills that you should develop during the course of your practical work.

(c) The angle of the track should be varied to give different values of θ (or $\tan \theta$). ✓

A graph of a against $\tan \theta$ should be plotted, ✓ which should be a straight line through the origin (sketch should show this). ✓

The gradient of this graph is g. ✓

The graph will reduce random errors by averaging several values. ✓

If there is a systematic error due to frictional forces, this will be indicated by a small intercept on the $\tan \theta$ axis, but the gradient will still be g. ✓

ⓔ It is important that you understand how a graph can reduce both random and systematic errors.

Knowledge check answers

Knowledge check answers

1 **a** $kg\,m^2\,s^{-2}$, **b** As, **c** $kg\,m^2\,s^{-3}A^{-2}$, **d** $kg\,m^2\,s^{-3}$
2 Diameter = $1.27 \times 10^{-2}\,m$
Length = $0.800\,m$
Area = $1.27 \times 10^{-4}\,m^2$
Volume = $1.02 \times 10^{-4}\,m^3$
3 Wavelength = $6.43 \times 10^{-7}\,m$
Slit distance = $3.3 \times 10^{-6}\,m$
4 $c_w = \dfrac{c}{n_w}$ and $c_o = \dfrac{c}{n_o}$ so $c_w : c_o = n_o : n_w$
$c_w : c_o = 1.47 : 1.33 = 1.11$
5 **a** 48, **b** 300, **c** 8.0×10^3
6 $n = \sin 35.5°/\sin 26.0° = 1.32$
7

Angle	$\sin\theta$	$\cos\theta$	$\tan\theta$
$\theta = 0°$	0	1	0
$\theta = 45°$	0.707	0.707	1
$\theta = \pi\,rad$	0	−1	0
$\theta = \pi/2\,rad$	1	0	∞

8 **a** 1.08×10^{21}, **b** 3.0×10^8, **c** 1.8×10^2 or 180
9 **a** 10^{18}, **b** 10^3, **c** 10^2, **d** 10^9
10 **a** Assume popcorn is cube of side $10^{-2}\,m$ or volume $10^{-6}\,m^3$
Assume room is cube of side $10\,m$ or volume $10^3\,m^3$
Number of pieces of popcorn = $10^3\,m^3/10^{-6}\,m^3 = 10^9$
b Assume you are made out of water. Number of electrons per molecule is 10^1
Mass of one molecule of water is $10^{-27}\,kg$
Assume your mass is $10^2\,kg$
Number of molecules of water you contain = $10^2/10^{-27} = 10^{29}$
$\Rightarrow$ Number of electrons = $10^{29} \times 10^1 = 10^{30}$
c $E_p = mg\Delta h$
Assume Mount Everest is $10^5\,m$ high:
$E_p = 10^2\,kg \times 10^1\,m\,s^{-2} \times 10^5\,m = 10^8\,J$
Assume energy from chocolate bar = $10^5\,J$
Number of bars = $10^8/10^5 = 10^3$
11 **a** $W = VQ$, **b** $x \propto F$, **c** $\phi = BA$
12 **a** $V = \dfrac{m}{\rho}$ **b** $m = \dfrac{\rho}{v}$
c $R_2 = R_1\left(\dfrac{V_{out}}{V_{in} - V_{out}}\right)$ **d** $a = \dfrac{v^2 - u^2}{2s}$
e $\theta = \sin^{-1}\left(\dfrac{n\lambda}{d}\right)$ **f** $\omega = \sqrt{\dfrac{F}{mr}}$
13 From $s = ut + \frac{1}{2}at^2 \Rightarrow t^2 = 2h/g$
(as $u = 0$) $\Rightarrow t = 0.55\,s$
14 **a** Moment = $225\,N \times 0.07\,m = 15.8\,N\,m$
b $L = x/\varepsilon = 0.10\,m/0.1 = 1\,m$
c $x = 0.040\,m \times \cos[(\pi/3)\,rad\,s^{-1} \times 11\,s]$
$x = 0.020\,m$ (remember to put your calculator in radian mode)

15 Acidity (pH)
Earthquake intensity (Richter scale)
Hardness (Mohs scale)
Star brightness (magnitude)
Wind intensity (Beaufort scale)
16 **a** plot λ (y-axis) against $\dfrac{1}{f}$ (x-axis); gradient = c
b plot R (y-axis) against l (x-axis); gradient = $\dfrac{\rho}{A}$
c plot T^2 (y-axis) against m (x-axis); gradient = $\dfrac{4\pi^2}{k}$
17 Instantaneous acceleration = $(20\,m\,s^{-1} - 5.5\,m\,s^{-1})/$
$(6.0\,s - 0.0\,s) = 2.4\,m\,s^{-2}$
Average acceleration = $17.5\,m\,s^{-1}/6.0\,s = 2.9\,m\,s^{-2}$
18 **a** Energy stored in a spring: $0.36\,J - 0.38\,J$
b Charge: $3300\,\mu C - 3400\,\mu C$
19 **a** graph a, **b** graph c, **c** graph d
20 Volume = $\pi r^2 h$
$= \pi \times (6.0 \times 10^{-3}\,m)^2 \times 1.50\,m = 1.7 \times 10^{-4}\,m^3$
21 Mass of one atom = molar mass/Avogadro constant
$= 12.0 \times 10^{-3}\,kg\,mol^{-1}/6.02 \times 10^{23}\,mol^{-1}$
$= 1.99 \times 10^{-26}\,kg$
Volume of one atom = mass/density
$= 5.69 \times 10^{-30}\,m^3$
Assuming atom is spherical:
$V = \dfrac{4}{3}\pi r^3$
$r = \sqrt[3]{\dfrac{3V}{4\pi}} = 1.1 \times 10^{-10}\,m$
22 **a** $0.262\,rad$, **b** $0.873\,rad$, **c** $2.53\,rad$
23 **a** $22.5°$, **b** $85.9°$, **c** $8.59°$
24

Angle	$\sin\theta$	$\cos\theta$	$\tan\theta$
$5.0\,rad$	−0.96	0.28	−3.4
$0.50\,rad$	0.48	0.88	0.55
$0.050\,rad$	0.050	1.0	0.050

25 There are many examples of random errors where measurements are affected in an unpredictable fashion: e.g. changes in room temperature, contact resistance, trying to read off the current from an ammeter when the current is changing quickly, human reaction time when starting and stopping a stopwatch.
26 $0.86 - 0.02 = 0.84\,mm$
27 Accurate — how close to the true value a measurement is.
Precision — how close to each other repeated measurements are.
28 The range of data is large ($62\,mT - 46\,mT = 16\,mT$), giving a 16% variation either side of the mean value. The data are therefore not precise.
29 The smallest change in value that can be measured using the instrument.

30 Largest = 100 + 5% = 105 Ω
Smallest = 100 − 5% = 95 Ω

31 a Uncertainty in measurement = 0.1 mm.
Diameter = 21.5 mm ± 0.1 mm

   b percentage uncertainty = (0.1 mm/21.5 mm) × 100% = 0.5%

32 Inconsistent decimal places in the first column. Incorrect column headings (should be pd/V and Current/A). Units should not appear in the body of the table. Current only recorded to 1 sf.

33 Mean a = 0.453 m s^{-2} = 0.45 m s^{-2} to 2 sf

34 a Correct reading is 8.3 cm (or 83 mm)

   b With a mirror behind the pointer, move your head until the image is hidden by the pointer. Then you are reading the scale vertically, which will give you the correct value.

35 Vernier reads 12.27 cm (or 122.7 mm)

36 Volume of sphere $V = \dfrac{4}{3}\pi r^3$

$V = \dfrac{4}{3} \times \pi \times \left(5.0 \times 10^{-3}\,\text{m}\right)^3 = 5.24 \times 10^{-7}\,\text{m}^3$

Density = mass/volume
= $(1.3 \times 10^{-3}\,\text{kg})/(5.24 \times 10^{-7}\,\text{m}^3)$ = 2483 kg m^{-3}
= 2.5×10^3 kg m^{-3} or 2.5 g cm^{-3} to 2 sf (in line with the data)

37 a 0.5 mm (the bottom half of the main scale subdivides the 1 mm divisions of the top half of the scale into 0.5 mm)

   b 7.25 mm, assuming the micrometer reads zero when closed/there is no zero error

38 There is a positive zero error, so this must be subtracted from each reading. The actual diameter is therefore: 2.12 mm − 0.03 mm = 2.09 mm.

39 Random errors must be reduced. You should be able to analyse each source of random error in the experiment and suggest ways of reducing it. This will include taking repeats, but will also involve choosing apparatus with greater resolution or improving experimental technique.

40 Use vernier callipers (or a micrometer). Check for zero error. Then measure the diameter several times, at different angles, and average.
Radius = ½ × diameter. Use the equation:
volume of a sphere $V = \dfrac{4}{3} \times \pi r^3$

41 a Human reaction time is at least 0.1 s, which is far more significant than the resolution of the stopclock (typically 0.01 s) when measuring short times.

   b Uncertainty = ±0.1 s (or possibly ±0.2 s)

42 (0.1 s/18.9 s) × 100% = 0.53%

43 Use the set-square with one side on the bench to check that the vertical height of the beam above the bench is the same in different places along the length of the beam.

44 a $\sin\theta$ = 87 mm/1000 mm = 0.087 $\Rightarrow$ θ = 5.0°

   b Uncertainty in lengths = (say) ±2 mm
% uc in height = (2 mm/87 mm) × 100% = 2.3%
% uc in length = (2 mm/1000 mm) × 100% = 0.2%
% uc in angle = 2.3% + 0.2% = 2.5%

45 $\frac{1}{2}mv^2 = mg\Delta h \Rightarrow v = \sqrt{2g\Delta h} \Rightarrow$
$v = \sqrt{2 \times 9.81 \times 0.10} = 1.4\,\text{m s}^{-1}$

46 (1.44 + 1.42 + 1.46 + 1.43)/4 = 1.44
1.56 is not included in the mean calculation, as it is clearly anomalous.

47 60 s × 0.5 s^{-1} = 30 measurements

48 Using $s = ut + \frac{1}{2}at^2$ where u = 0 and a = 9.81 m s^{-2}:
$t = \sqrt{\dfrac{2s}{a}} = \sqrt{\dfrac{2 \times 0.05\,\text{m}}{9.81\,\text{m s}^{-2}}} = 0.101\,\text{s} \approx 100\,\text{ms}$

49 120 m = k × 2000 kg
x = k × 1400 kg
x = (1400 kg/2000 kg) × 120 m = 84 m

50 a Plot v^2 on y-axis against h on x-axis. The graph should be a straight line through the origin with gradient $2g$.

   b Plot T^2 on y-axis against l on x-axis. The graph should be a straight line through the origin with gradient $4\pi^2/g$.

51 $R = V/I = (4.97 \times 10^{-3}\,\text{V})/(22.6 \times 10^{-6}\,\text{A})$ = 220 Ω

52 a For a metallic conductor, the potential difference is directly proportional to the current, provided the temperature remains constant.

   b Your sketch should have axes labelled V/V and I/A and show a straight line through the origin.

   c If the component is ohmic, the graph will be a straight line through the origin.

   d If V is plotted against I, the resistance will be equal to the gradient.

53 ln(8.2) − ln(8.0) = 2.10 − 2.08 = 0.02
ln(8.4) − ln(8.2) = 2.13 − 2.10 = 0.03
The point ln(8.2) = 2.10 should be plotted with an error bar between 2.08 and 2.13

54 Graph should be a straight line of negative slope. The gradient is equal to −λ (where λ is the decay constant).
Then $t_{1/2} = \dfrac{\ln 2}{\lambda} = \dfrac{0.693}{\lambda}$.

55 a $\%\,\text{difference} = \dfrac{(28.3 - 26.7)\,\text{N m}^{-1}}{27.5\,\text{N m}^{-1}} \times 100\% = 6\%$

   b $\%\,\text{difference} = \dfrac{(9.81 - 9.71)\,\text{m s}^{-2}}{9.81\,\text{m s}^{-2}} \times 100\% = 1\%$

56 a $\ln T = \ln\left(\dfrac{2\pi}{\sqrt{k}}\right) + \dfrac{1}{2}\ln m$

   b Gradient = ½ = 0.50

   c Intercept = $\ln\left(\dfrac{2\pi}{\sqrt{k/\text{N m}^{-1}}}\right) = 0.19$

Knowledge check answers

$$\Rightarrow \frac{2\pi}{\sqrt{k/\text{Nm}^{-1}}} = e^{0.19} = 1.21$$

$$k = \left(\frac{2\pi}{1.21}\right)^2 = 27\,\text{Nm}^{-1}$$

57 $f = \dfrac{1}{2l}\sqrt{\dfrac{T}{\mu}}$

$$f = \frac{1}{2 \times 0.50\,\text{m}}\sqrt{\frac{10\,\text{N}}{2 \times 10^{-3}\,\text{kg}\,\text{m}^{-1}}}$$

$$= 70.7\,\text{Hz} \approx 70\,\text{Hz}$$

58 $\dfrac{1}{\text{N}^{-1}\text{s}^{-2} \times \text{m}^2} = \text{N}\,\text{s}^2\,\text{m}^{-2}$

$$= \text{kg}\,\text{m}\,\text{s}^{-2} \times \text{s}^2\,\text{m}^{-2} = \text{kg}\,\text{m}^{-1}$$

59 $n\lambda = d\sin\theta$

$$\Rightarrow \sin\theta = \frac{n\lambda}{d} = \frac{1 \times 550 \times 10^{-9}\,\text{m}}{\left(300 \times 10^3\,\text{m}^{-1}\right)^{-1}} = 0.165$$

$$\theta = \sin^{-1}(0.165) = 9.5° \approx 10°$$

60 a The distances would be greater (as the wavelength is longer).

b $n = \dfrac{d\sin\theta}{\lambda}$ and $\sin\theta \leq 1$

$$n \leq \frac{\left(300 \times 10^3\,\text{m}^{-1}\right)^{-1} \times 1}{693 \times 10^{-9}\,\text{m}} \leq 4.8$$

n must be an integer, so $n_{\text{max}} = 4$

61 a The waves emitted from each source have the same frequency and a constant phase relationship.

b The waves emitted from the two slits originate from the same wavefront. As all points on a wavefront are in phase, the waves emitted by the slits must be coherent.

c From $w = \dfrac{\lambda D}{s} \Rightarrow \lambda = \dfrac{ws}{D}$

$$\lambda = \frac{1.0 \times 10^{-3}\,\text{m} \times 0.50 \times 10^{-3}\,\text{m}}{0.80\,\text{m}}$$

$$= 6.25 \times 10^{-7}\,\text{m} = 625\,\text{nm}$$

62 Using $s = ut + \frac{1}{2}at^2$

$1.2\,\text{m} = 0 + \frac{1}{2} \times 9.8\,\text{m}\,\text{s}^{-2} \times t^2$

$t = 0.49\,\text{s} \approx 0.5\,\text{s}$

63 $E = \dfrac{Fl}{A\Delta l} \Rightarrow \Delta l = \dfrac{Fl}{AE}$

$$\Delta l = \frac{5.0\,\text{N} \times 3.0\,\text{m}}{\pi \times \left(0.188 \times 10^{-3}\,\text{m}\right)^2 \times 1.2 \times 10^{11}\,\text{Nm}^{-2}}$$

$$= 1.1 \times 10^{-3}\,\text{m} = 1\,\text{mm}$$

64 Stress = F/A

$$= \frac{2.00\,\text{kg} \times 9.8\,\text{Nkg}^{-1}}{\pi \times \left(0.19 \times 10^{-3}\,\text{m}\right)^2} = 1.73 \times 10^8\,\text{Pa}$$

Strain = $\Delta l/l$

$$= \frac{4.1 \times 10^{-3}\,\text{m}}{2.94\,\text{m}} = 1.39 \times 10^{-3}$$

$$E = \frac{\text{stress}}{\text{strain}} = \frac{1.73 \times 10^8\,\text{Pa}}{1.39 \times 10^{-3}} = 1.2 \times 10^{11}\,\text{Pa}$$

65 Units for E are $\dfrac{\text{Nm}^{-1} \times \text{m}}{\text{m}^2} = \text{N}\,\text{m}^{-2} = \text{Pa}$

66 $R = \dfrac{\rho l}{A} = \dfrac{1.08 \times 10^{-6}\,\Omega\text{m} \times 0.50\,\text{m}}{\pi \times \left(\frac{1}{2} \times 0.315 \times 10^{-3}\,\text{m}\right)^2}$

$R = 6.93\,\Omega \approx 7\,\Omega$

67 $R = \dfrac{\rho l}{A} \Rightarrow A = \dfrac{\rho l}{R} \Rightarrow \dfrac{\pi d^2}{4} = \dfrac{\rho l}{R}$

$$\Rightarrow d = \sqrt{\frac{4\rho l}{\pi R}}$$

$$d = \sqrt{\frac{4 \times 1.08 \times 10^{-6}\,\Omega\text{m} \times 1.00\,\text{m}}{\pi \times 18.3\,\Omega}}$$

$d = 2.74 \times 10^{-4}\,\text{m}$ (or $0.274\,\text{mm}$)

68 From $V = \varepsilon - Ir$, when $I = 0$ we have voltage = $\varepsilon = 1.58\,\text{V}$. Connecting the resistor across the terminals creates a current I in the cell, so the $1.58\,\text{V}$ drops by Ir to $1.50\,\text{V}$.

$$I = \frac{\varepsilon}{R+r} \text{ and } I = \frac{V}{R}$$

$$\Rightarrow \frac{\varepsilon}{V} = 1 + \frac{r}{R} \Rightarrow r = R \times \left(\frac{\varepsilon}{V} - 1\right)$$

$$r = 4.7\,\Omega \times \left(\frac{1.58\,\text{V}}{1.50\,\text{V}} - 1\right) = 0.25\,\Omega$$

69 $V = \varepsilon - Ir$

$1.70\,\text{V} = 2.68\,\text{V} - (1.20 \times 10^{-3}\,\text{A}) \times r$

$r = 817\,\Omega \approx 800\,\Omega$

70 Squaring $T = 2\pi\sqrt{\dfrac{m}{k}}$ on both sides gives

$$T^2 = \frac{4\pi^2 m}{k} = \frac{4\pi^2}{k} \times m$$

71 From the above answer, a graph of T^2 against m will have a gradient of $4\pi^2/k$. So:

$$\text{gradient} = \frac{(0.550 - 0.00)\,\text{s}^2}{(0.325 - 0.00)\,\text{kg}} = 1.69\,\text{s}^2\,\text{kg}^{-1}$$

$$k = \frac{4\pi^2}{\text{gradient}} = \frac{4\pi^2}{1.69\,\text{s}^2\,\text{kg}^{-1}} = 23.3\,\text{kg}\,\text{s}^{-2}$$

72 a PE_{max} at the top of the swing (maximum displacement). It momentarily has zero velocity as its direction reverses.

b KE_{max} at the bottom of the swing, when it is moving fastest (zero displacement).

73 From $T = 2\pi\sqrt{\dfrac{l}{g}} \Rightarrow l = \dfrac{T^2 g}{4\pi^2}$

$$l = \frac{(1.00\,\text{s})^2 \times 9.81\,\text{m}\,\text{s}^{-2}}{4\pi^2} = 0.248\,\text{m}$$

74 $\sin 20° = 0.3420$ and $20° = 0.3490\,\text{rad}$

% difference $= \dfrac{0.3490 - 0.3420}{0.3455} \times 100\% = 2.03\% = 2\%$

75

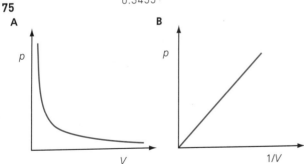

76

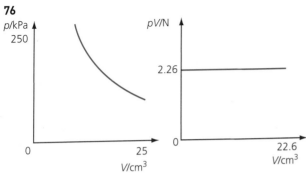

Units of pV are $\text{Pa}\,\text{m}^3 = \text{N}\,\text{m}^{-2}\,\text{m}^3 = \text{N}\,\text{m} = \text{J}$

77 a An ideal gas is a model of a gas that obeys the ideal gas equation at all temperatures. However, real gases *condense to a liquid* as they cool.

b Absolute zero is the temperature at which the molecules of matter are considered to have no kinetic energy (in quantum mechanics, the molecules have a minimum kinetic energy, called the end-point energy). The accepted value of the zero of the Kelvin temperature scale is −273.15°C.

c As the pressure is constant:

from $\dfrac{V_1}{T_1} = \dfrac{V_2}{T_2} \Rightarrow V_2 = \dfrac{V_1 T_2}{T_1}$

$V_2 = \dfrac{4.0\,\text{cm}^3 \times (177 + 273)\,\text{K}}{(27 + 273)\,\text{K}} = 6.0\,\text{cm}^3$

78 a $t = RC$

$t = 3.3 \times 10^3\,\Omega \times 220 \times 10^{-6}\,\text{F} = 0.73\,\text{s} \sim 1\,\text{s}$

b In the equation $V = V_0 e^{-t/RC}$, t/RC must be just a number, without any unit. Thus RC must have the same unit as t, i.e. second.

c $R = \dfrac{V}{I}$ and $C = \dfrac{Q}{V}$, so

$RC = \dfrac{V}{I} \times \dfrac{Q}{V} = \dfrac{Q}{I} = \dfrac{C}{Cs^{-1}} = s$

79 $I = V/R = 5.8\,\text{V}/(3.34 \times 10^3\,\Omega) = 1.7\,\text{mA}$

80 Fleming's left-hand rule can be used to determine the direction of the force on a current-carrying wire placed in a magnetic field.

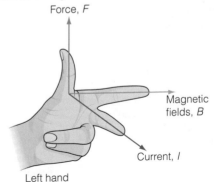

If the first finger points in the direction of the field, and the second finger points in the direction of the current, then the thumb gives the direction of the force on the wire.

81 $F = BIl = 0.11\,\text{T} \times 1.0\,\text{A} \times 45 \times 10^{-3}\,\text{m}$
$= 4.95 \times 10^{-3}\,\text{N} \approx 5\,\text{mN}$

82 Angle between *axis* of coil and field $\theta = (90 - 30)° = 60°$
$A = \pi \times (6.0 \times 10^{-3}\,\text{m})^2 = 1.13 \times 10^{-4}\,\text{m}^2$
Flux linkage $= 0.20\,\text{T} \times 1.13 \times 10^{-4}\,\text{m}^2 \times 500\,\text{turns} \times \cos 60°$
$= 5.7\,\text{mWb-turns}$

83 From Faraday's law, an emf is induced in a coil if the magnetic flux linking the coil changes. As the alternating current in the large coil, and therefore the magnetic field it produces, is continuously changing, the flux linking the search coil is continuously changing. An alternating emf is thus induced in the search coil.

84 The *y-gain* controls the distance the dot moves vertically up or down the screen per volt. The settings can vary from microvolts per division (μV/div) to volts per division (V/div).

The *time base* controls the distance the dot moves horizontally on the screen per second. Typically, settings for the time base are in milliseconds per division (ms/div).

85 As $I \propto \dfrac{1}{d^2} \Rightarrow I_1 d_1^2 = I_2 d_2^2$

a $160 \times 1.00^2 = 40 \times d_2^2$
$d_2 = 2.0\,\text{m}$

b $160 \times 1.00^2 = 640 \times d_2^2$
$d_2 = 0.50\,\text{m}$

86 As $I \propto \dfrac{1}{d^2}$,

$\dfrac{I_{4m}}{I_{1m}} = \dfrac{d_{1m}^2}{d_{4m}^2} = \dfrac{1^2}{4^2} = \dfrac{1}{16}$

Index